BUCKINGHAMSHIRE
COUNTY COUNCIL

This book has been
withdrawn from the
County Library stock

Price: 20p

MAMMALS
OF BRITAIN AND EUROPE

Iain Bishop

Illustrated by Bernard Robinson

KINGFISHER

First published in 1982 by Kingfisher Books Limited
Elsley Court, 20–22 Great Titchfield Street, London W1P 7AD

© Kingfisher Books Limited 1982

All Rights Reserved

BRITISH LIBRARY CATALOGUING IN PUBLICATION DATA
Bishop, Iain
  Mammals.– (Kingfisher Guides; 17)
  1. Mammals
  I. Title
  599 QL706.2
  ISBN 0 86272 026 5

Designed and edited by Eric Inglefield
Colour separations by Newsele Litho, Milan, London
Phototypeset by Southern Positives and Negatives (SPAN),
Lingfield, Surrey
Printed and bound in Italy by Vallardi Industrie
Grafiche, Milan

CONTENTS

Introduction	8
Marsupials	14
Primates	15
Insectivores	16
Hedgehogs	16
Moles	17
Shrews	18
Bats	22
Horse-shoe Bats	24
Vespertilionid Bats	26
Free-tailed Bats	31
Rabbits, Hares and Pikas	32
Rabbits and Hares	32
Rodents	34
Squirrels	34
Beavers	40
Porcupines	41
Coypus	42
Dormice	43
Hamsters, Voles, Rats and Mice	46
Birch Mice	63
Carnivores	64
Bears	65
Dogs and allies	66
Weasels and allies	70
Mongooses and Genets	78
Raccoons	79
Cats	80
Seals, Walruses and Sea Lions	82
Seals	82
Walruses	86
Odd-toed hooved mammals	88
Horses	88
Even-toed hooved mammals	89
Pigs	89
Cattle, Sheep, Goats and allies	90
Deer	96
Whales, Dolphins and Porpoises	104
Rorquals and Humpback Whales	105
Right Whales	108
Sperm Whales	110
Beaked Whales	111
White Whales	113
Dolphins	114
Porpoises	119
Glossary	120
Index	121

INTRODUCTION

Mammals are *vertebrate* animals—that is, they have backbones—and they are called *warm-blooded* because they can to some extent maintain a warm body temperature independent of the outside temperature. They feed their young on milk produced in the mammary glands of the mother. Most mammals have a covering of hair, although there are a few exceptions. Whales, for example, have only a few bristles on the snout and chin of young animals. There are about 4,000 *species*, or different kinds, of mammals in the world, but only about 200 of these can be found in Europe, and some of these are rare whilst others have been introduced by man.

The range of mammals living in Europe, known as its *fauna*, has been moulded by two major historical factors. First, a succession of Ice Ages at different times rendered large parts of the continent uninhabitable. The last Ice Age was of particular importance as it covered Europe with a sheet of ice extending from the North Pole to a line from southern Britain eastwards to the south of Berlin, Warsaw and Moscow. At that time, some 15,000 years ago, even the sunny Mediterranean countries were cold and inhospitable. As the ice retreated, the plants and animals were able to establish themselves further and further north, and others came from northern Africa and Asia so that our present European fauna has many relatives in those regions.

The other important factor in shaping Europe's fauna was the arrival, spread and (more recently) the rapid increase in the human population. Settlement, the growth of agriculture, industry and forestry have profoundly affected the landscape and vegetation of the area and hence the habitats available for the animals to live in. Man has also had an even more direct impact on the mammals. Hunting for food, fur and sport has affected different species in different ways, both on land and at sea. The breeding and husbandry of domestic animals and the introduction of non-native species, either deliberately for pleasure or commerce, or accidentally, have altered the composition of Europe's mammal life.

The Habitats of Europe

From the Arctic wastes surrounding the North Pole to the warmth of Mediterranean countries there are a number of climatic zones in Europe which are reflected in the major vegetation types which can grow in the region. These zones are irregular and modified by local conditions, such as the soil, climate, water and altitude. The distribution of these zones is shown in the map. The range and habitat of each species of mammal described in this book are given in the text, but few are restricted to a single habitat and some are found throughout the continent. However, each vegetation zone does have a number of characteristic mammals which can normally be expected to occur there. The tundra zone supports a population of Musk oxen and this is also the home of the reindeer in summer.

A red deer stag in the highlands of Scotland.

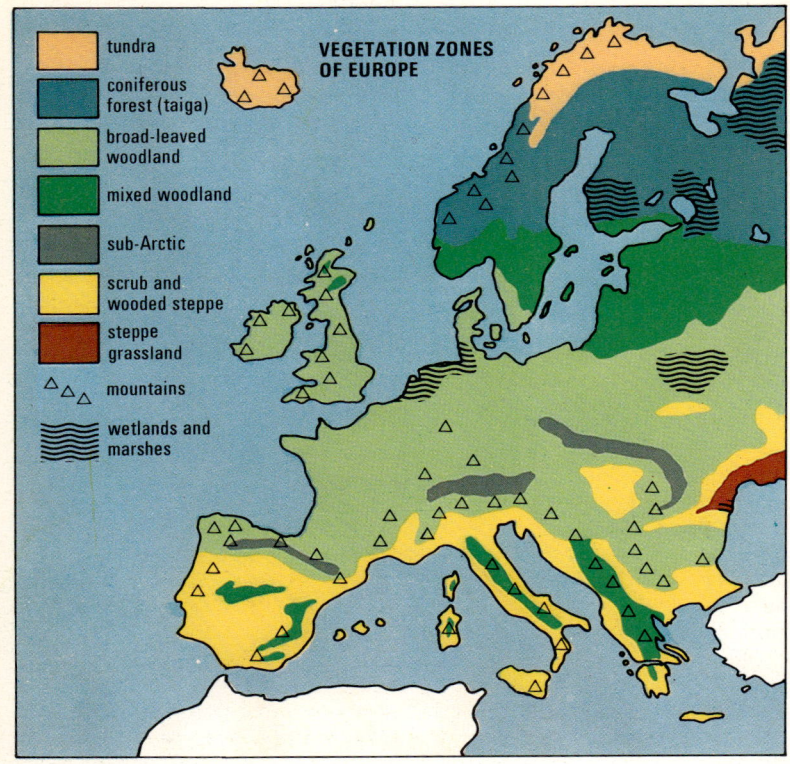

Amongst the smaller tundra animals are the Arctic fox, lemming and a number of voles. In the extensive and less rigorous taiga habitat, Pine marten, sable, Red squirrel and more mice and voles occur. A large part of central Europe is covered with mixed or broadleaved forest, which contains a rich fauna including large numbers of bats, shrews, squirrels, other rodents and some of our most familiar animals such as Red deer, Roe deer, foxes, Wild cat and bison. The extensive areas of scrub and grassland are the preferred habitat of hedgehogs, shrews, many rodents, jackals, polecats, rabbits and hares. Amongst the most widespread species are the Common mole, Striped Field mouse, wolves, and Roe deer.

The habitat zones of Europe have been much modified by man, and the reduced extent of forests in some areas is of particular concern, for many of our native species of mammals are primarily forest animals. The legal protection afforded to mammal species is not, and cannot be, effective if the protected animals have no habitat in which to thrive. The fundamental importance of habitat conservation cannot be too strongly emphasized if our wildlife is to continue to thrive.

The Skulls and Teeth of Mammals

Walking in the countryside one can often find the bones and teeth of mammals lying on the ground, turned up by a plough or washed up on the beach. The wide range of mammals illustrated in this book, which includes such dissimilar animals as reindeer and shrews, bats and whales, is reflected in the diversity of their skeletal structure. A study of these bone structures is important not only for the identification of an animal and its classification but also to discover how an animal's bone structure is related to its way of life. The form of the teeth is particularly interesting from this point of view.

Teeth are set in sockets in the jaws and are described in four main types depending on their position, shape and function. Most mammals develop a set of 'milk teeth' which are present during the period of rapid development of the jaw and are replaced by the larger permanent teeth when jaw growth begins to slow down. The permanent *incisor* teeth are usually small, flattened, single-rooted teeth set at the front of the jaws and used for nibbling, gnawing and biting. The *canines* (eye teeth or dog teeth) are larger, also single-rooted and set behind the incisors. They are used to slash, and hold large prey, and only a single canine is present in each jaw above and below. The cheek teeth have more than one root and may be more complex in shape; they are the *premolars* and *molars* set towards the back of the jaws. The premolars are variable in number and are preceded by milk teeth during development; the molars develop only once. These teeth are used for chewing, cutting or grinding food into suitable-sized pieces for swallowing and digesting.

Not all these kinds of teeth are found in all animals, but the basic arrangement is three incisors, one canine, four premolars and three molars. This number is found in the pig, but most other species have fewer teeth. The teeth of insectivores and most bats cannot be clearly divided into the standard types, and are very sharp, with lots of *cusps*, or points, to seize, hold and chew their prey. Rodents have a remarkable arrangement of teeth quite unlike any other group of mammals. They have only a single incisor tooth on each side of each jaw followed by a long gap known as a *diastema*; they have no canines, and the cheek teeth, usually three or four, are broad and used to grind and chew the vegetable matter that most species eat. Carnivores usually have large canine teeth, or *fangs*, and the teeth behind these are sharp and blade-like and used for cutting or shearing flesh and bone with an action not unlike that of scissors. These teeth are known as *carnassials*. Seals have many simple teeth used to grasp fish and squid and not divided into different types. The teeth of hooved mammals are very varied, but the pattern is generally similar throughout. The upper incisors are often missing, being replaced by a horny pad against which the lower incisors can bite. Canines are also often missing, there is a distinct diastema, and the cheek teeth are broad, ridged or cusped to grind hard plant food. Some whales have peg-like teeth, varying in number from one to many, whilst others have completely lost all the teeth and evolved a system for filtering food from the sea.

The general form of a mammal's skull is also related to the way in which it feeds, particularly to the size and placement of the jaw muscles. Other parts of the body are similarly modified from a general pattern to reflect the way of life of a particular species. Limbs of deer and horses, adapted to swift movement, can be contrasted with those of moles, modified for digging, of the wallaby, developed for hopping, and the extreme example of the whale, whose limbs have been evolved into flippers for locomotion in water. When examining the bones of animals remember that they have a function, and that this can often be detected in their form.

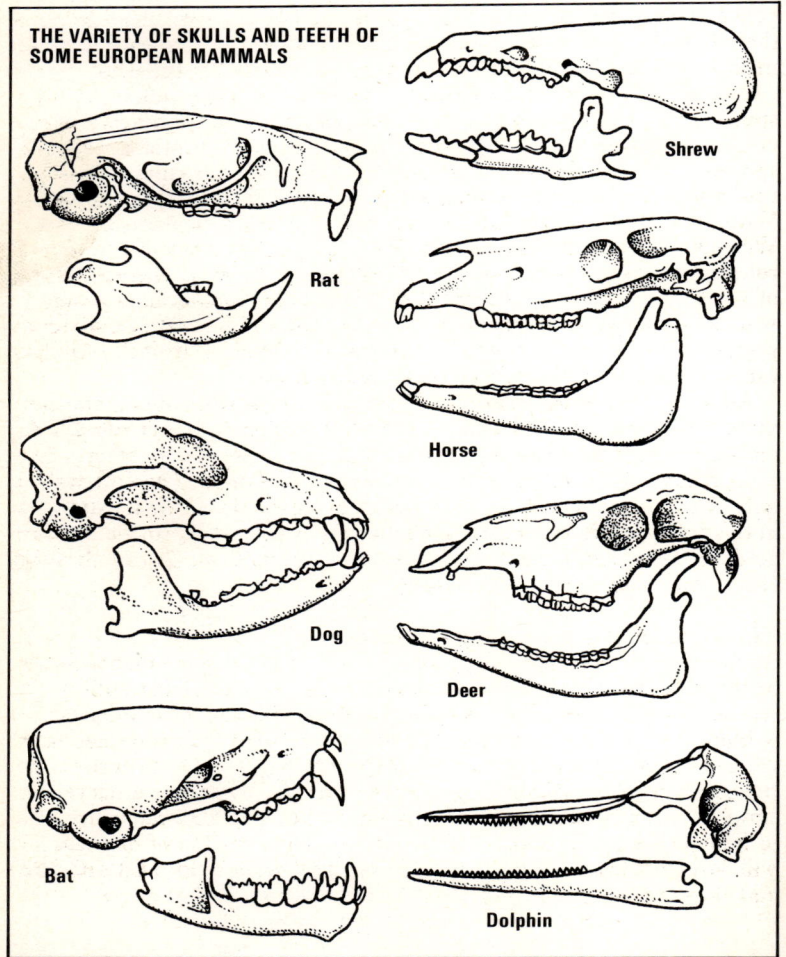

THE VARIETY OF SKULLS AND TEETH OF SOME EUROPEAN MAMMALS

Rat

Shrew

Horse

Dog

Deer

Bat

Dolphin

The Classification and Identification of Mammals

Scientists have divided the world's living mammals into 18 major groups known as *orders*, each with a Latin name. Eleven of these groups are represented in Europe and are included in this book although not all are native to the region. Within each order similar animals are grouped in *families*, such as the Mole family (Talpidae), which belongs to the Order Insectivora, or insect eaters. Each member of a family, or *species*, is given a scientific or Latin name consisting of two parts—the *generic* name and the *specific* name. For example, *Talpa europaea*, *Talpa caeca* and *Talpa romana* are all moles belonging to a group of similar animals called the genus *Talpa*, but their various specific names indicate that they are thought to be distinct or at least sufficiently different from each other to make it unlikely that they would normally interbreed in the wild.

As the study of mammals develops, scientists' opinions about whether species are 'good' (whether they are sufficiently distinct to deserve different names) change from time to time, and the names used for a particular species may also change according to a set of agreed rules. Thus the reader may find the same animal with a different name in different books. Within the broad groups of orders there are numerous sub-divisions. Weasels and stoats are similar small carnivores adapted to subtly different roles in the ecology of Europe. Even within a single species there is a range of variation which is often confusing. There are changes in appearance with season, differences due to age, to the food and habitat of individual populations and to the genetic make-up of individual animals. All this variability often leads to difficulties of identification.

Added to the above problems which confront the student of mammals, there is the practical difficulty of finding and seeing the animals. Mammals are secretive, many species are nocturnal and, apart from a few exceptionally bold and large species, they are seldom seen, and often then only as a glimpse as they scuttle away in the vegetation. The illustrations in this book (adult males unless labelled otherwise) will provide a useful guide to the identification of European mammals which can then be checked against the descriptive text.

The Conservation of Europe's Mammals

As in other parts of the world man has often been in conflict with the mammal fauna of Europe, and usually it is the fauna that suffers. In recent years, however, there is an increasing awareness of the value of our wildlife and its place in the quality of our lives. All over Europe laws have been passed to protect the more endangered species either to ensure that the commercially valuable species can continue to provide a harvest or simply in recognition of the richness which their presence can contribute to human lives. The main impetus for the conservation of our wildlife will, however, continue to depend upon the individual naturalist's interest and enthusiasm.

MARSUPIALS
Order Marsupialia

This order includes about 250 species found mainly in Australasia and America, particularly South America. They are characterized by their method of reproduction, in which the gestation period is short and the tiny young are born in a very under-developed state. They then crawl to the pouch (*marsupium*), where they are protected and nourished for a long period. In some species there is no pouch, and the young remain attached to the nipple. Animals in this order are adapted to a wide variety of life styles; kangaroos and wallabies (Family Macropodidae) have a similar life style to the deer.

RED-NECKED WALLABY
Macropus rufogriseus

Head and body 60–70cm, tail 60–70cm. Somewhat larger than the hare, but easily recognized by its long, stout, tapering tail, short ears and characteristic bounding gait, using its long, strong hind legs and feet. Colour variable, but usually a grizzled grey-brown above, whitish below. Tail grey with a black tip. Black also on paws, feet, muzzle and ears.

Range and habitat: a native of Australia and Tasmania, and common in parks and zoos. Some escapes from parks in Britain have led to the establishment of 2 feral populations. The wallaby normally lives in woodland and scrub.

Food and habits: grazes and browses on grasses and shrubs, including heather. Active in daytime. Lives in small groups. Survives well in captivity, but severe winters may reduce feral populations drastically.

Breeding: a single young is carried in the pouch for about 9 months. In Britain most young are born in spring, but births can occur at any time.

Red-necked Wallaby

female

PRIMATES
Order Primates

The order includes the lemurs, monkeys, apes and man. About 180 species inhabit the warmer regions of Asia, Africa and the Americas. The most important characteristic is the well-developed brain enclosed in a large, rounded brain-case. There are usually 5 digits on each limb, and the thumb and forefinger are opposable and used for gripping.

MONKEYS
Family Cercopithecidae

This family contains all the African and Asian monkeys, about 60 species, including the so-called Barbary 'ape', which is really a monkey. All the monkeys, except the baboons, share a short muzzle, forwardly directed eyes and grasping hands and feet. Nearly all are diurnal and omnivorous.

BARBARY APE
Macaca sylvanus
Head and body 60–70cm; the only member of the family without a tail. The only monkey found in Europe, having been introduced from North Africa.
Range and habitat: Rock of Gibraltar, where about 40 live on the ground in rocky woodlands.
Food: diet of insects, fruit and leaves is supplemented by feeding.
Breeding: a single young is born in summer.

INSECTIVORES
Order Insectivora

This widely distributed order includes some 300 species in 8 families. In Europe 3 families are found, the hedgehogs (Erinaceidae), moles and desmans (Talpidae) and shrews (Soricidae). All are smallish, long-snouted, short-legged animals with sharp, cusped teeth and 5 clawed digits on each limb. They are mostly nocturnal and eat a variety of animal food, including insects and small vertebrates.

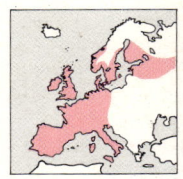
Western Hedgehog

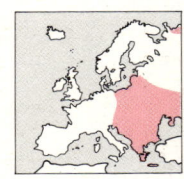
Easte n Hedgehog

Western Hedgehog

HEDGEHOGS
Family Erinaceidae

Hedgehogs are covered with barbless spines, with hair on the face, limbs and underside. They can roll into a tight ball when disturbed. They burrow, but can also swim and climb well. They are found in Europe, Asia, Africa and some islands in habitats as varied as forest and desert. There are 3 European species.

Eastern Hedgehog

WESTERN HEDGEHOG
Erinaceus europaeus
Head and body 19–27cm. Underside a uniform dull yellowish-brown, with a narrow parting of the head spines.

EASTERN HEDGEHOG
Erinaceus concolor
Similar to the Western Hedgehog, but has a clear white patch on the breast.

ALGERIAN HEDGEHOG
Erinaceus algirus
Very pale below, and has a wide parting. Found only in southern areas, to which it was probably introduced from North Africa.

MOLES
Family Talpidae

A family of about 20 species found in northern Eurasia and North America. It includes moles, which live underground in tunnel systems dug with spade-like forefeet, and some aquatic or semi-aquatic forms such as the desmans. Most species are insectivorous.

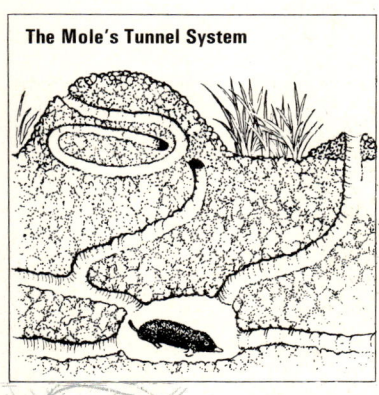

The Mole's Tunnel System

NORTHERN MOLE
Talpa europaea
Head and body 12–15cm, tail about 40mm, hind foot 17–19mm. Dark grey to black fur, which can lie in either direction. Active day and night. Produces a single litter of 3–4 young in spring.

BLIND MOLE
Talpa caeca
Smaller than other European moles (hind foot 16–17mm), and its eyes are permanently closed by a membrane. Restricted to southern Europe.

ROMAN MOLE
Talpa romana
Also has permanently closed eyes, but is larger than the Blind mole, with the hind feet over 17·5mm. Found in southern Europe.

Northern Mole

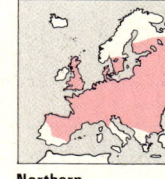

Northern Mole

PYRENEAN DESMAN
Galemys pyrenaicus
Head and body 11–13·5cm, tail 13–15cm. Muzzle long, flat and broad; tail flattened from side to side and fringed with stiff hairs; hind feet large and webbed.
Range and habitat: holes in the banks of mountain streams and canals in northern Spain and Portugal.
Food: small aquatic invertebrates.
Breeding: 1 litter of 4 young born in spring.

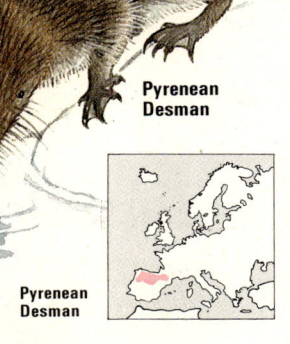

Pyrenean Desman

Pyrenean Desman

SHREWS
Family Soricidae

This large family contains some 200 species. Shrews are small, short-legged animals with long pointed noses. Their fur is short and usually brown or grey, they have very small eyes and their senses of hearing and smell are well developed. Most species are terrestrial, active day and night and solitary. They are usually carnivorous or insectivorous, though a little plant food is eaten. Because the many species are difficult to identify, a detailed examination of teeth and skulls is often necessary. The 15 European species can be split into 2 main groups: the red-toothed shrews (*Sorex* and *Neomys*) and the white-toothed shrews (*Suncus* and *Crocidura*).

COMMON SHREW
Sorex araneus
Head and body 70–85mm, tail 45mm, hind foot 12–13mm. The teeth are red-tipped. The silky fur on the back is dark brown, the underside pale, and along the sides there is a distinct band of an intermediate colour.
Range and habitat: Britain, Scandinavia and eastern Europe, wherever the ground cover is sufficient: grass hedgerows, woodland and moorland.
Food and habits: eats mainly small soil animals, insects, worms and spiders; will also eat carrion. Solitary life style, and active periodically throughout day and night.
Breeding: 3 or more litters produced in May–September, each with 5–8 young. Life span up to 15 months.

MILLET'S SHREW
Sorex coronatus
This species is almost indistinguishable from the Common shrew. Small differences in skull proportions and chromosomes are used to identify it.
Range: Northern Spain, France, the Netherlands, parts of Germany and Switzerland.
Habits: similar to Common shrew.

SPANISH SHREW
Sorex granarius
Although slightly smaller than the Common shrew, this species is very difficult to identify without an examination of the skull and chromosomes.
Range and habitat: the mountains of central Spain.

APENNINE SHREW
Sorex samniticus
Similar to the Common shrew, but its tail is shorter (less than 40mm).
Range: southern Italy.

PYGMY SHREW
Sorex minutus
Head and body 45–60mm, hind foot usually under 12mm, tail relatively long. Distinctly smaller than the Common shrew and no band of intermediate colour is present on the flanks.
Range: widespread in Europe, but found only at high altitude in south.

LAXMANN'S SHREW
Sorex caecutiens
This species is identified by its teeth.
Range and habitat: forest and tundra of Scandinavia and Poland.

LEAST SHREW
Sorex minutissimus
Head and body 35–44mm, hind feet 8–9mm. A very small shrew.
Range: Scandinavia, Finland and east to Japan.

DUSKY SHREW
Sorex sinalis
Similar in size to the Common shrew, but the underside is almost the same colour as the upper.
Range: Scandinavia and Finland east to Asia.

ALPINE SHREW
Sorex alpinus
Larger than the Common shrew and its tail is markedly longer, being equal to its head and body length.
Range and habitat: found mainly in coniferous forest in the mountains of central Europe, the Pyrenees and the Balkans.

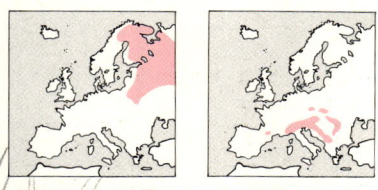

Laxmann's Shrew **Alpine Shrew**

Laxmann's Shrew

Alpine Shrew

Water Shrew

Miller's Water Shrew

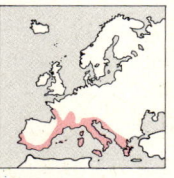
Pygmy White-toothed Shrew

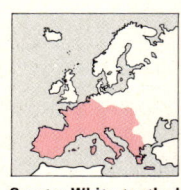
Greater White-toothed Shrew

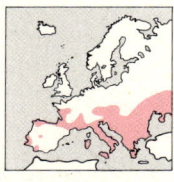
Lesser White-toothed Shrew

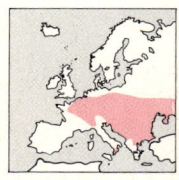
Bicoloured White-toothed Shrew

The Water shrew hunts on land as well as in water. It uses its poisonous saliva to paralyse prey.

WATER SHREW
Neomys fodiens
Head and body 70–90mm, hind foot 17–20mm. A large, dark shrew, nearly black above, with underside colour varying from silver-grey to black. The teeth are red-tipped. Fringes of stiff hairs on the underside of the tail and hind feet assist swimming.
Range and habitat: found throughout Europe, except Ireland, in dense vegetation near fresh water.
Food and habits: insect larvae and other water invertebrates. Also uses its poisonous saliva to kill larger prey, such as small fishes, frogs and even small mammals.

Breeding: at least 3 litters every year, each with 5–8 young.

MILLER'S WATER SHREW
Neomys anomalus
This slightly smaller shrew has a reduced tail fringe, which occurs only on the end third of the tail.
Range and habitat: distribution patchy, but found with the Water shrew in some places.

Water Shrew

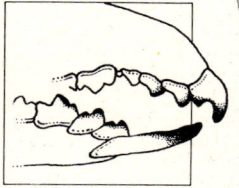
Teeth of Water shrew

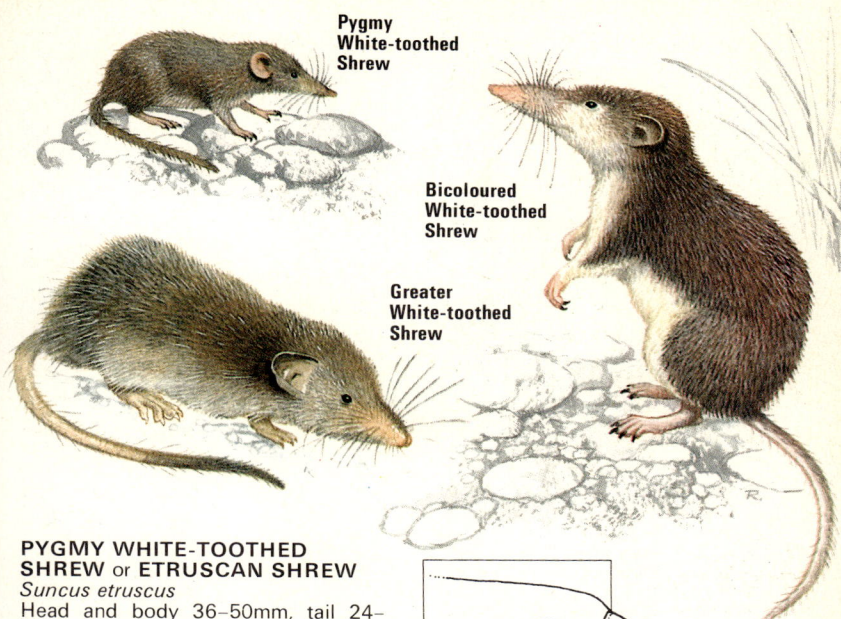

PYGMY WHITE-TOOTHED SHREW or ETRUSCAN SHREW
Suncus etruscus
Head and body 36–50mm, tail 24–29mm, hind foot 7–8mm. The smallest mammal in the world. Usually weighing less than 2g. Long hairs project above normal fur and form whiskers on the tail. Reddish-grey above, dull grey below. Teeth without any trace of red pigment.

Range and habitat: widespread in grassland and often found in arable farmland and gardens.

Food and habits: little is known about its habits, but in captivity it is known to eat large insects.

GREATER WHITE-TOOTHED SHREW
Crocidura russula
Head and body 65–85mm, tail 35–50mm, hind foot 12–13mm. Has whiskers on the tail, and is greyish-brown above, fading gradually to a yellowish-grey below.

Range and habitat: dry grassland, woodland and hedges in southern and central Europe.

Habits: similar to Common shrew. Active day and night. Preyed upon by barn owls.

Breeding: young born from February to November in southern Europe.

Teeth of Greater White-toothed shrew

BICOLOURED WHITE-TOOTHED SHREW
Crocidura leucodon
Head and body 64–87mm, tail 28–39mm, hind foot 12–13mm. The dark grey upper side contrasts sharply with the yellowish-white below. Tail whiskered and clearly bicoloured.

Range and habitat: often found on the fringes of woodland and the steppes of eastern Europe.

LESSER WHITE-TOOTHED SHREW
Crocidura suaveolens
Very similar to *Crocidura russula*, but somewhat smaller. Head and body 55–70mm, tail 30–40mm, hind foot 11–12mm. Tail whiskered. Certain identification depends on detailed examination of teeth.

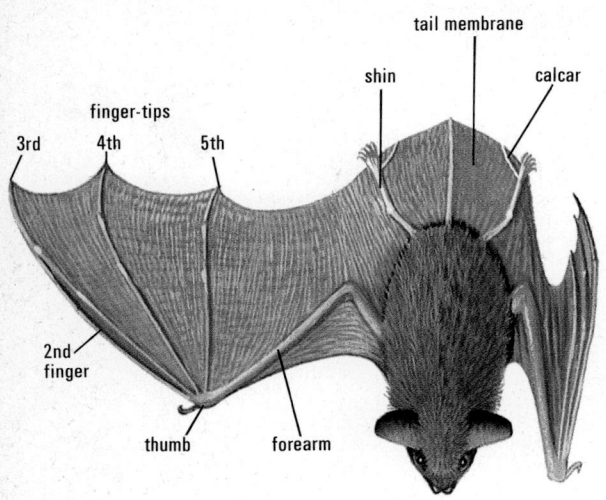

The External Physical Structure of a Bat (a Pipistrelle)

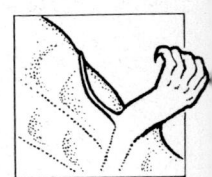

The calcar, a spur of cartilage supporting the tail membrane.

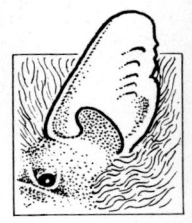

The tragus, a pointed lobe at the base of the ear.

BATS
Order Chiroptera

Bats are the only mammals that fly, although several species of other orders can glide. Their wings are skin membranes supported by fingers, fore and hind limbs and tail. About 1,000 species are currently recognized. Bats are distributed throughout the world, with the exception of the polar regions, but most species are tropical or subtropical. Fruit bats (Sub-order Megachiroptera) are not found in Europe, but some 30 species of insectivorous bats (Sub-order Microchiroptera) have been recorded from the area. Bats usually fly at night using high-pitched sound pulses to navigate and find food. They spend the daylight hours in a torpid state in dark, sheltered cavities in trees, caves, attics or cellars. In temperate regions most bats hibernate in winter. Correct identification of some European bats is difficult and may depend upon the minute detail of structure of the skull or teeth, but most genera and some species can be determined by an examination of the external characters shown in the diagram. Three families of bats are found in Europe: Horse-shoe bats (Rhinolophidae), Vespertilionid bats (Vespertilionidae), and Free-tailed bats (Molossidae).

The Common pipistrelle is the most abundant European bat and is often found in very large colonies, especially in its winter roosts.

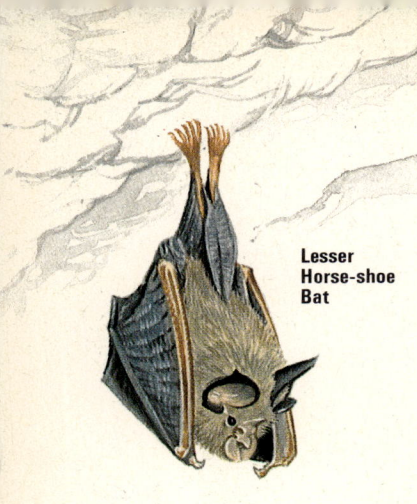

Lesser Horse-shoe Bat

Greater Horse-shoe Bat

HORSE-SHOE BATS
Family Rhinolophidae

Members of this family, which has 5 European species, are easily recognized by the presence of complex skin lobes on the face. Horse-shoe bats are colonial and rest in clusters in caves, where they hang freely from the roof, usually with their wings wrapped tightly around the body. They hibernate from October to March. Mating occurs in summer, but fertilization is delayed as the females can store sperm. There is usually a single young.

LESSER HORSE-SHOE BAT
Rhinolophus hipposideros
The smallest Horse-shoe bat in Europe. Head and body about 63mm, forearm 35–40mm. Dark grey above, lighter below. Ears large and broad. The lobes on the face are associated with the emission of ultra-sound through the nose. Large summer colonies in attics and buildings disperse to form smaller groups in caves and cellars in winter. The single young clings to its mother's fur until too large to be carried.

GREATER HORSE-SHOE BAT
Rhinolophus ferrumequinum
The largest Horse-shoe bat in Europe. Head and body about 70mm, forearm 50–60mm. The thick fur is grey above, pale buff below. Ears large and broad with curved pointed tips.

MEDITERRANEAN HORSE-SHOE BAT
Rhinolophus euryale
Head and body up to 58mm, forearm 43–49mm. Blue-grey above, creamy below. Slender point above sella.

BLASIUS'S HORSE-SHOE BAT
Rhinolophus blasii
Forearm 44–50mm. Similar to Mediterranean Horse-shoe bat, but with a broad nose-leaf and a very constricted sella.

MEHELY'S HORSE-SHOE BAT
Rhinolophus mehelyi
Forearm 48–55mm. Similar in size to the Greater Horse-shoe bat, but much paler in colour and with a sharp point above sella.

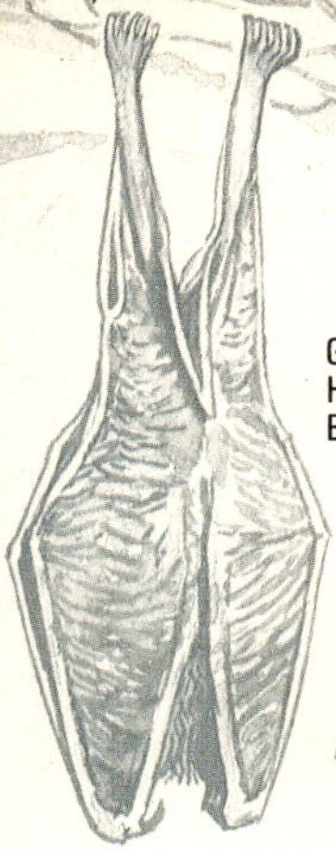

Greater Horse-shoe Bat at rest

Mediterranean Horse-shoe Bat

Mehely's Horse-shoe Bat

Blasius's Horse-shoe Bat

Nose-leaves of Horse-shoe bats (front view)

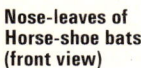
— lancet
— sella
Lesser Horse-shoe bat

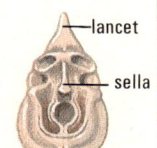

Greater Horse-shoe bat   Mediterranean bat

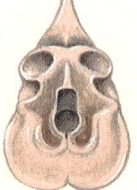

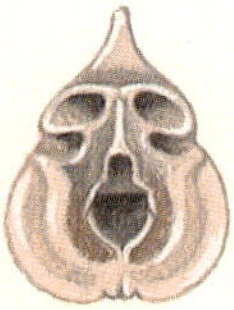

Blasius's bat    Mehely's bat

Lesser Horse-shoe Bat

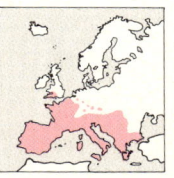

Greater Horse-shoe Bat

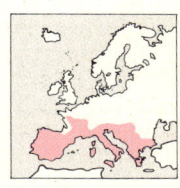

Mediterranean Horse-shoe Bat

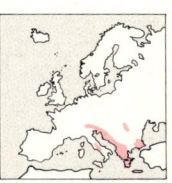

Blasius's Horse-shoe Bat

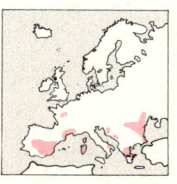

Mehely's Horse-shoe Bat

VESPERTILIONID BATS
Family Vespertilionidae

This family of small bats includes about 280 species, of which 26 are European. The family is widely distributed in tropical and temperate areas. The nose has no complex lobes, and the tail never projects beyond the membrane by more than one or two vertebrae. Three groups can be recognized within the European members of the family: the Myotis bats, which are mostly small, with a long, slender, pointed tragus in the ear; the larger noctules and serotines, with a short, kidney-shaped tragus; and the small pipistrelles, with a short, blunt tragus. The pipistrelles, noctules and serotines have a lobe of tail membrane extending outside the calcar which is absent in Myotis bats.

MYOTIS BATS

DAUBENTON'S BAT
Myotis daubentoni
A small bat; head and body 50mm, forearm about 35mm. Very large feet more than half the length of the shin. The tragus is straight and slender. Greyish-red above, dull white below. The ears and muzzle pinkish-brown.
Habitat: lives in wooded country often near water, roosting in trees and buildings in summer, but sheltering in caves in winter to hibernate.

NATHALINE BAT
Myotis nathalinae
Similar to Daubenton's bat, but slightly larger; forearm 33–36mm. Difficult to distinguish without microscopic examination.
Range: distribution is not well known, although it has been recorded from Spain, France and Switzerland.

LONG-FINGERED BAT
Myotis capaccinii
Similar to Daubenton's bat, but has distinctly more hair on the shin and adjacent membrane, and is somewhat paler in colour.

POND BAT
Myotis dasycneme
A larger species; head and body about 59mm, forearm 45mm. Upper parts yellowish-brown, pale brown-grey below.

BRANDT'S BAT
Myotis brandti
Very small; forearm 32–37mm. Reddish-brown above, buff below, with very dark, almost black membranes.

Natterer's Bat

Daubenton's Bat

Brandt's Bat

WHISKERED BAT
Myotis mystacinus
Another small species; head and body about 50mm, forearm 32-37mm. Dark grey fur.

GEOFFROY'S BAT
Myotis emarginatus
Forearm 36-42mm. Similar to the Whiskered bat, but with reddish fur, and its ears are deeply notched on the hind margin. The wing membrane meets the hind limb at the base of the outer toe.

NATTERER'S BAT
Myotis nattereri
Similar to preceding species, but the ear notch is much less distinct and there is a fringe of hair on the membrane near the tail. Tragus very long and pointed.

Whiskered Bat

BECHSTEIN'S BAT
Myotis bechsteini
A larger bat; forearm 39-44mm. Very long ears. Greyish-brown. The last tail vertebra projects beyond tail membrane.

Daubenton's Bat

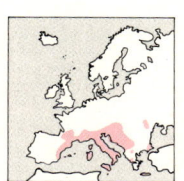
Long-fingered Bat

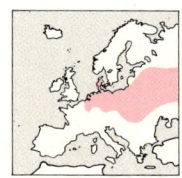
Pond Bat

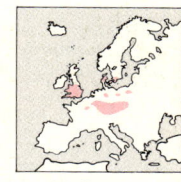
Brandt's Bat

Whiskered Bat

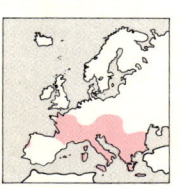
Geoffroy's Bat

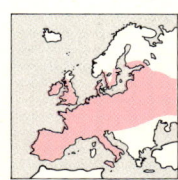
Natterer's Bat

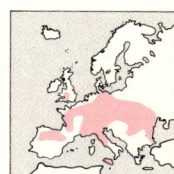
Bechstein's Bat

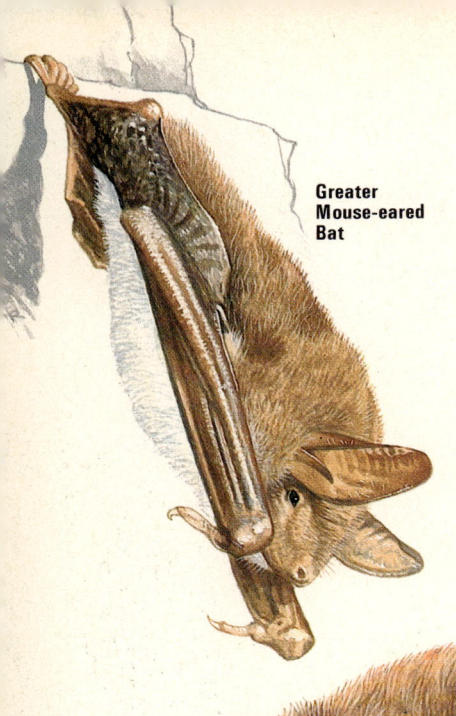

Greater Mouse-eared Bat

NOCTULES, SEROTINES AND RELATED BATS

NOCTULE BAT
Nyctalus noctula
Large; head and body about 82mm, forearm 46–55mm. Fur rich golden brown above and below. Noctules are colonial bats roosting in trees, seldom in caves. They fly high in search of insects and may migrate over considerable distances.

GREATER NOCTULE BAT
Nyctalus lasiopterus
Similar to the Noctule bat, but distinctly larger; forearm 62–69mm. Very rare.

LEISLER'S BAT
Nyctalus leisleri
Also known as the Lesser Noctule, this species is smaller than the Noctule; head and body about 63mm, forearm 40–45mm. It is also darker in colour.

Lesser Mouse-eared Bat

GREATER MOUSE-EARED BAT
Myotis myotis
The largest of the Myotis bats; head and body 65–90mm, forearm 57–67mm. Broad wings, very large ears, narrow pointed tragus.

LESSER MOUSE-EARED BAT
Myotis blythi
Large, with head and body up to 74mm, but forearm is usually less than 59mm. Slender tragus and narrow muzzle.

Greater Mouse-eared Bat

Noctule Bat

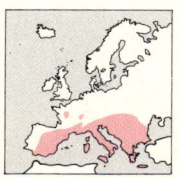

Lesser Mouse-eared Bat

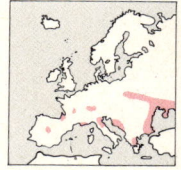

Greater Noctule Bat

SEROTINE BAT
Eptesicus serotinus
Head and body 76mm, forearm 48–55mm. Dull brown above and below. Tail tip extends beyond membrane by about $1\frac{1}{2}$ vertebrae.

NORTHERN BAT
Eptesicus nilssoni
Smaller than the Serotine and lighter in colour; forearm 36–46mm. Ranges further north than any other European bat.

PARTI-COLOURED BAT
Vespertilio murinus
Head and body 55–63mm, forearm 40–48mm. Similar to the Serotine, but dark brown above, with white tips to hairs giving a frosted appearance, and pale below. Tail projects one vertebra beyond membrane.

Noctule Bat

Leisler's Bat

Serotine Bat

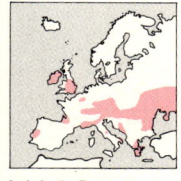

Leisler's Bat

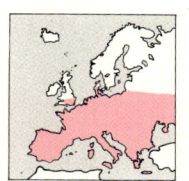

Serotine Bat

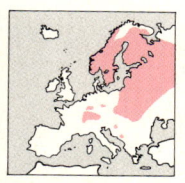

Northern Bat

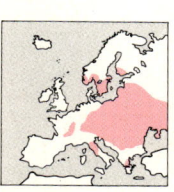

Parti-coloured Bat

Northern Bat

PIPISTRELLES AND ALLIES

COMMON PIPISTRELLE
Pipistrellus pipistrellus
The smallest European bat; head and body 38-50mm, forearm 27-32mm. Ear short, broad and triangular, tragus short and blunt. Fur variable in colour from dark brown to light reddish, but uniform. All Pipistrelles are difficult to identify without a detailed examination of teeth.
Habitat: forms large colonies in trees and buildings, sheltering in caves in winter.
Food: insects are caught and eaten in flight.

NATHUSIUS'S PIPISTRELLE
Pipistrellus nathusii
Similar to the Common Pipistrelle, but is less uniform in colour and is slightly larger; forearm 32-36mm.

KUHL'S PIPISTRELLE
Pipistrellus kuhli
Somewhat lighter in colour, with a distinct white margin on the membrane between the last digit and the foot.

SAVI'S PIPISTRELLE
Pipistrellus savii
Darker above than below, and the light-tipped hairs have dark bases.

COMMON LONG-EARED BAT
Plecotus auritus
Head and body about 50mm, forearm 34-41mm. Distinguished by extremely long ears, the bases of which meet on top of the head. Tragus long and narrow. Can hover to take insects off tree leaves.

GREY LONG-EARED BAT
Plecotus austriacus
Similar to the Common Long-eared bat, but the upper fur is greyer in colour and the tragus is somewhat broader. Averages a slightly longer forearm, 38-43mm.

BARBASTELLE
Barbastella barbastellus
Small; head and body about 50mm, forearm 37-42mm. Has short, broad ears with bases meeting on top of head. Snout short and dark. Colour dark, almost black, with some white hair tips on lower back.

SCHREIBER'S BAT
Miniopterus schreibersi
Medium-sized; forearm 42-48mm. Grey-brown above, paler below. The long tail is completely enclosed in membrane. Very short muzzle.

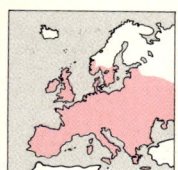
Common Pipistrelle

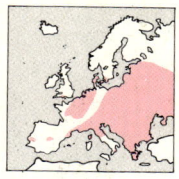
Nathusius's Pipistrelle

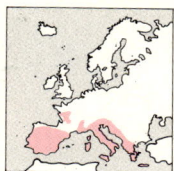
Kuhl's Pipistrelle

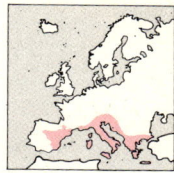
Savi's Pipistrelle

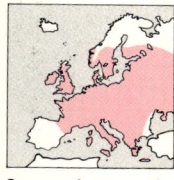
Common Long-eared Bat

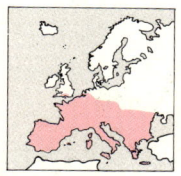
Grey Long-eared Bat

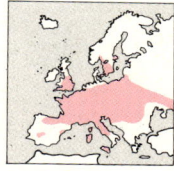
Barbastelle

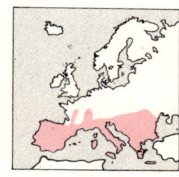
Schreiber's Bat

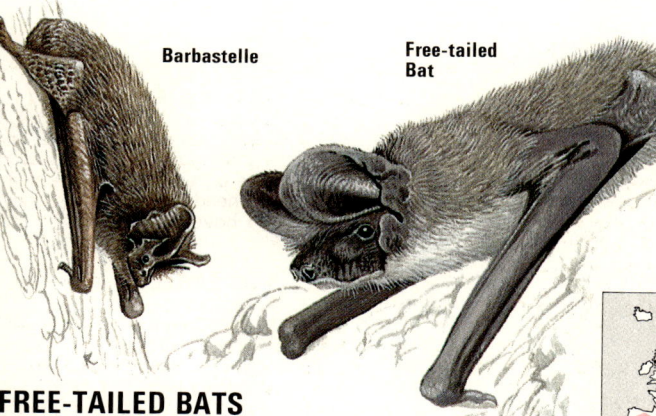
Barbastelle

Free-tailed Bat

Free-tailed Bat

FREE-TAILED BATS
Family Molossidae

A very large family found mostly in the tropics. Of the 80 species, only one is found in Europe. The family is easily recognized by the appearance of the tail, which projects far beyond the free edge of the rather narrow tail membrane. Even in Europe the bats seldom hibernate, but remain active throughout the year.

FREE-TAILED BAT
Tadarida teniotis
Head and body up to 85mm, forearm 58–64mm; tail up to 55mm, of which about half is free of the membrane. The only bat in Europe with the tail extending more than 2 vertebrae beyond the membrane. It often lives in rocky areas and flies high and fast.

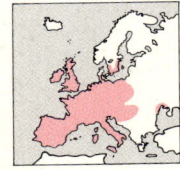
Rabbit

warren

RABBITS, HARES AND PIKAS
Order Lagomorpha

This order contains a number of small- to medium-sized animals which are terrestrial and vegetarian, and have a small tail or none. They all have 2 pairs of upper incisors set one behind the other. Despite some superficial resemblance they are not related to the rodents.

RABBITS AND HARES
Family Leporidae

Members of this family are easily recognized by their long ears, long hind legs and short well-furred tails. The feet have hairy pads. The sensory systems of hearing, sight and smell are well developed, and an ability to run very fast on strong hind legs affords protection. Many of the 50 species are useful for food and fur, but they are also, in large numbers, pests of agriculture.

Rabbit

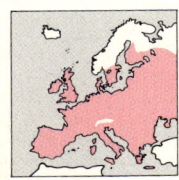
Brown Hare

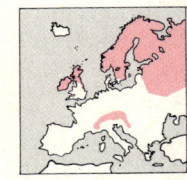
Mountain Hare

RABBIT
Oryctolagus cuniculus
Head and body up to 40cm, hind foot 7·5–9·5cm. Differs from the hare in its small size, shorter ears and legs. Colour very variable from brown to sandy, and black specimens are common in some areas. Eyes large and prominent.
Range and habitat: widespread in grassland, especially with cover and in sand dunes.
Food and habits: lives in colonies in underground warrens. Active mainly at night. Feeds mainly on grass, but can destroy cereal crops, roots and young trees. Prey of weasels, foxes, cats, dogs and birds of prey, and is hunted for food and fur by humans.
Breeding: litters of 4–6 young are born underground in spring and autumn.

BROWN HARE
Lepus capensis
Head and body 50–60cm, hind foot 12–15cm. Yellowish-brown above, white on belly. Has long ears with black tips, and black and white tail.
Habitat: mainly a lowland species, found especially on farmland.
Food and habits: feeds on bark, grain and other vegetation, and often becomes a pest of agriculture and forestry. Solitary, active mainly at night. Prey of fox, wild cat and eagle, and hunted by humans.
Breeding: usually 2 young (*leverets*) are born above ground in long grass, and there are 3–4 litters each year. Courtship involves spectacular leaping, chasing and boxing.

MOUNTAIN HARE
Lepus timidus
Head and body 50–60cm, hind foot 12–15cm. More heavily built than Brown hare, with shorter ears and longer legs. Tail white above and below. Summer coat brown with a blue tinge. Often turns white in winter except for ear tips. In far north may remain white all year round, but in some localities, such as Ireland, it never turns white.
Habitat: lives in mountainous areas above tree line in grassland and moorland.
Food: grasses and shrubs such as heather.

summer coat

winter coat

Mountain Hare

RODENTS
Order Rodentia

This is the largest order of mammals, with about 1,500 species distributed throughout the world. About 60 species are represented in Europe. They occupy all the available habitat types and have evolved to a variety of life styles, including burrowing, tree climbing, gliding and swimming. All the rodents have 4 sharp, chisel-shaped, continuously growing incisor teeth for gnawing, behind which is a gap, or *diastema*, before the grinding or chewing cheek teeth (usually 3 or 4 in each jaw). The incisors are kept sharp because the hard enamel surface at the front wears less rapidly than the softer dentine behind. Nearly all species are small, although a few, such as the beaver, agouti and capybara, reach medium size. The distribution of many species has been modified by some form of relationship with human beings. The accidental introduction of pest species, the deliberate introduction of food species, and escapes from fur farms and the pet trade have all helped to spread rodents around the world. Several species are major pests of agriculture, forestry and food storage; others are involved in the transmission of diseases of man and domestic animals. Some are used in medical research.

SQUIRRELS
Family Sciuridae

This family includes not only the familiar tree squirrels but also the ground squirrels and some 'flying' (gliding) squirrels. There are about 350 species widely distributed, but only 7 occur in Europe. Almost all eat nuts, seeds, and other plant material. Most species are diurnal, but the flying squirrels are nocturnal. Several species are regarded as pests, others as a source of fur.

Red Squirrel

summer coat

Red Squirrel

spring coat

Red Squirrel

RED SQUIRREL
Sciurus vulgaris
Head and body 22cm, tail 18cm. The only native tree squirrel in Europe. Colour very variable from red forms to black, with seasonal variation especially pronounced in northern areas. Summer coat usually a rich red with a darker mid-back area; winter coat greyish. In summer, dark ear tufts are prominent, but these are moulted in winter. The Red squirrel is trapped for its fur principally in eastern Europe.
Range and habitat: lives throughout the wooded areas of Europe, especially in coniferous forest.
Food and habits: active in daytime, feeding mainly on seeds and nuts, but also eats young tree shoots. Does not hibernate, but may sleep for long periods in winter. The Red squirrel usually has several sleeping nests, and the breeding nest may be in a hollow tree. Prey of Pine and Beech martens, stoats and birds of prey.
Breeding: breeding season varies with locality, but there are usually 2 litters per year with 3-4 young in each.

winter coat

GREY SQUIRREL
Sciurus carolinensis
Head and body 26cm, tail 21cm. Winter coat grey with yellowish-brown mid-dorsal stripe. In summer, coat has reddish tinge, especially on the sides. Tail has light-coloured fringes, and ear tufts are lacking.
Range and habitat: originating in the woodlands of North America, it has been introduced several times to Britain and Ireland, where it is now widespread in deciduous woodlands. It is also common in the parks and gardens of southern Britain.
Food: can kill trees by stripping bark for food, but main diet is nuts, particularly acorns and a wide range of roots, shoots, bulbs and even some insects, eggs, and small birds.

FLYING SQUIRREL
Pteromys volans
Head and body 16cm, tail 12cm. Upper parts grey or grey-brown, underside white. Large eyes, small ears. It glides by using a furred membrane extended between the fore and hind limbs, and partly supported by a spur of cartilage on each wrist.
Range and habitat: lives in the forests of northern Russia and Finland.
Food and habits: eats seeds (acorns, chestnuts and pine seeds) and also fruit and small birds. Climbs up trees then glides for up to 50m. Nocturnal, it hides in a hollow tree by day. Hibernates. Prey of martens and owls.
Breeding: usually 2 litters of 3–6 young are born each year.

The Grey squirrel is a serious pest of forestry but an attractive animal in town parks.

EUROPEAN SOUSLIK
Citellus citellus
Head and body 19–22cm, tail 5·5–7·5cm. A ground squirrel with short legs. The coarse fur is of a plain greyish-brown colour sometimes with indistinct lighter spotting. The underside is white or yellowish.
Range and habitat: dry Balkan steppe country, meadows and fields.
Food and habits: eats seeds, nuts, roots and leaves, and may also take insects, small mammals, birds and eggs. Stores food in autumn and fattens before hibernation. Active in daytime. Makes complex tunnel systems and lives in colonies.
Breeding: usually produces 1 litter with 2–13 young each year.

SPOTTED SOUSLIK
Citellus suslicus
Similar to European souslik but tail shorter, 30–40mm. Coat distinctly white-spotted. Also similar in habits, and can be a pest of agriculture.

Spotted Souslik

SIBERIAN CHIPMUNK
Tamias sibiricus
Head and body 15–22cm, tail 10–14cm. Easily recognized by the dark longitudinal stripes on a reddish-brown background; the underside is white or buff.
Range and habitat: introduced from northern Russia and Siberia to Greece, Germany, the Netherlands and Austria. Lives in wooded country with thick undergrowth.
Food and habits: eats fruit, nuts and seeds. Does not hibernate, but stores food in autumn and sleeps much of the winter.
Breeding: usually 6 young in each litter, with 2 litters per year.

European Souslik

burrow

Siberian Chipmunk

ALPINE MARMOT
Marmota marmota
Head and body 50–60cm, tail 13–16cm. The largest of the ground squirrels, it is heavily built, and has a broad head and small ears. It has reddish fur with grey on the head, shoulder and rump. The end of the tail is black.
Range and habitat: high mountains, where it burrows deeply in open alpine pastures above the tree line.
Food and habitats: eats grasses and sedges. Forms small family groups active during the day. Preyed on by foxes and eagles, but is very alert, and the colony employs a well-developed warning system. Hibernates October–April.
Breeding: a single litter of 2–4 young born each year.

Alpine Marmot

Alpine Marmot

burrow

BEAVERS
Family Castoridae

The family contains only one genus with 2 species. It was once widely distributed around the northern part of the world from Europe and Asia to North America, but hunting has reduced the populations to remnants in many places. American and Eurasian beavers are very similar in appearance, and some scientists suggest that they are the same species. However, recent studies on skull structure and chromosomes support the opinion that they are separate.

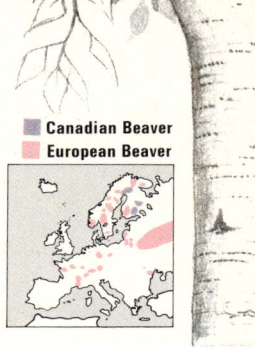

Canadian Beaver
European Beaver

EUROPEAN BEAVER
Castor fiber
Head and body 90cm, tail 35cm. The largest of the European rodents. Coat brown with very dense underfur and a sleek covering of guard hairs. Blunt-nosed, heavily built. Tail broad and flat with prominent scales. Short legs with webbed hind feet.
Range and habitat: exploited for fur and, although once widespread in northern Europe, it is now found only rarely in the Rhône and Elbe valleys, parts of Scandinavia, and Russia. Lives in rivers and lakes in wooded country.
Food and habits: feeds on a variety of aquatic vegetation and tree bark (aspen and willow are preferred). Will fell trees to reach tender shoots. The beaver is mainly nocturnal and spends much time in water. It lives in family groups of up to 12 animals in burrows in river banks or *lodges* built in the water. The burrows have several entrances, some under water, and are now more common than lodges, which are built of mud and sticks with a central chamber above water level. These constructions can dam streams, forming beaver ponds.
Breeding: usually has 1 litter per year, with 2-4 young.

CANADIAN BEAVER
Castor canadensis
Difficult to distinguish from the European beaver, but has a darker head and consistent skull differences. Introduced to Finland, where it is now well established.

European Beaver

North African Crested Porcupine

PORCUPINES
Family Hystricidae

The family includes all the Old World porcupines, comprising 12 species. They are large rodents living on the ground in small groups. The characteristic spines are not barbed and may reach 40cm in length. They are erected by a complex of muscles in the skin and can be rattled as a warning. When attacked, the animal can rush backwards at the enemy to impale it on the quills, which are easily detached.

NORTH AFRICAN CRESTED PORCUPINE
Hystrix cristata
Head and body 70cm, tail 10cm. A large brownish-black rodent with white band under neck. Head and neck with crest of long white and brown bristles. Body quills of two types, long and slender or short and stout, banded black and white. On rump, quills mainly black; on tail, white.
Range and habitat: probably introduced to southern Europe from North Africa. Lives in rocky hill country with cover.
Food and habits: roots, leaves, and bark. Nocturnal. Pest of root crops and forestry.
Breeding: 1 litter of 2–4 young born in spring underground in a burrow.

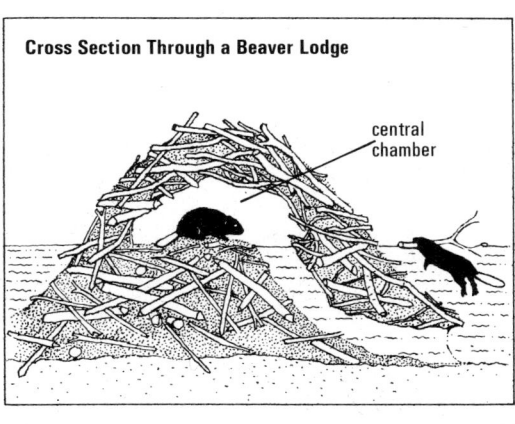

Cross Section Through a Beaver Lodge

central chamber

COYPUS
Family Capromyidae

A small family of about a dozen species which includes the hutia of the West Indies and the coypu of southern South America. They are all thick-bodied, short-legged animals. Several of the species of West Indian hutias have become extinct in recent historic time.

COYPU
Myocastor coypus
Head and body 60cm, tail 40cm, thick and scaly. Head large with short ears. Hind feet webbed. An aquatic animal native to southern South America but kept in captivity for fur farming in many parts of the world. Fur known as *nutria* is obtained by removing coarse guard hairs from pelt, leaving soft, thick underfur.
Range and habitat: escapes have populated several localities in Europe (West Germany, Holland, the Camargue and Loire valley in France, and East Anglia in Britain).
Food: water vegetation and shellfish.
Breeding: 2 litters are born each year with up to 9 young in each. Mammae are placed well up the side of body to permit suckling in water.

The characteristic chisel-like incisor teeth of rodents are well shown by the coypu, which is amongst the largest of the rodents. Coypu were introduced, or escaped from fur farms in many parts of Europe. They are a pest in some areas.

Coypu

Garden Dormouse

Garden Dormouse

DORMICE
Family Gliridae

A small family of about 10 species in Europe, Asia and Africa. Five species are found in Europe. The dormice resemble small squirrels. Most species have bushy tails, large eyes, rounded ears and short legs. They are climbing animals, eating mainly fruit buds and nuts. They are nocturnal and secretive and can be pests of orchards. They remain dormant for long periods in winter.

GARDEN DORMOUSE
Eliomys quercinus
Head and body 15cm, tail 9cm. The long slender tail has a flattened, black and white tuft at the tip. Body greyish-brown above, white below. Black mask on the face.
Habitat: woodland with shrubs and cultivated areas.
Food and habits: spends much of its time on the ground. Eats a wide variety of food including insects, snails, birds' eggs and nestling small mammals and birds, as well as fruit and nuts. Becomes dormant in autumn after the summer fattening period. It is nocturnal, hiding in a hole during day.
Breeding: 1 or 2 litters are born each year with 3-6 young in each.

43

FOREST DORMOUSE
Dryomys nitedula
Head and body 9–13cm, tail 9 cm. Greyish or yellowish-brown above, white below. Black face mask. Bushy tail without black and white tuft.
Habitat: deciduous woodland with shrub layer.
Food and habits: eats fruit, seeds and insects. Nocturnal.
Breeding: usually only 1 litter of 3–6 young each year.

FAT DORMOUSE
Glis glis
Head and body 14–19cm, tail 13cm. The largest dormouse. Grey-brown above, somewhat lighter below. Dark stripes on outside of legs, dark ring round eyes. Long bushy tail.
Habitat: woodland and orchards.
Food and habits: eats nuts, seeds, fruit and insects. Can be pest of fruit and forestry crops. Becomes very fat in autumn and dormant October–April. Nocturnal. In Roman times fattened for the table and considered a delicacy.
Breeding: usually 1 litter of 4–5 young each year.

COMMON DORMOUSE
Muscardinus avellanarius
Head and body 7cm, tail 7cm. A small dormouse, light tawny above, yellowish-white below. Throat and chest white. Tail thickly haired.
Habitat: deciduous forest with shrubs, copses and thick hedgerows.
Food and habits: eats nuts, seeds and some insects. Nocturnal. Sleeps from October to April rolled in a ball in a sheltered nest.
Breeding: often 2 litters per year with 3–4 young in each.

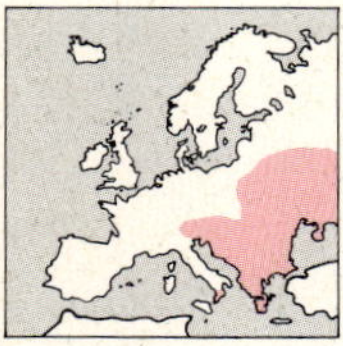

Forest Dormouse

Forest Dormouse

Fat Dormouse

Common Dormouse

Fat Dormouse

The Common dormouse often makes its summer nest from the stripped bark of honeysuckle mixed with grass and moss.

MOUSE-TAILED DORMOUSE
Myomimus roachi
Head and body 7–10·5cm, tail 6–8cm. Upper parts yellowish-grey, white below. Long tail with short hairs and no bush.
Habitat: dry woodland and scrub.
Habits: a rare species which is seldom encountered. Little is known about its habits.

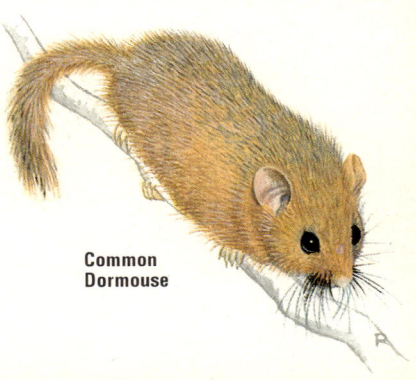

Common Dormouse

HAMSTERS, VOLES, RATS AND MICE
Family Muridae

This is the largest rodent family, containing over 1,000 species of rats, mice, voles, hamsters and their allies. Usually small, fast-breeding, short-lived animals, they are found in almost every corner of the world. All have 3 molar teeth in each jaw. The family is divided into several sub-families, two of which—the hamsters (Cricetinae) and the rats and mice (Murinae)—are often considered to be separate families in their own right. Many species are pests of agriculture and stored products and several carry diseases. Over 40 species are found in Europe, many of which are difficult to identify.

HAMSTERS AND ALLIES
Sub-family Cricetinae

A widely distributed sub-family of over 500 species, but with only 3 representatives in Europe. They are mostly small, terrestrial animals with simple grinding surfaces on the molars. In many parts of their range they closely resemble rats and mice, but the hamsters, with their short tails, small ears, and cheek pouches, are the distinctive European members of the family.

COMMON HAMSTER
Cricetus cricetus
Head and body 22–32cm, tail 3–6cm. The largest hamster. Above, the thick fur is light brown; below, it is black with white patches on the sides. Broad feet and large cheek pouches.
Habitat: grassland, meadows and cultivated land usually near water.
Food and habits: feeds mainly on seeds and grain, but also eats roots and insects. Makes extensive burrow systems with many chambers, some of which are used for food storage. Mainly nocturnal.
Breeding: usually 2 litters each year with 6–12 young. May be a serious pest of agriculture.

Common Hamster

RUMANIAN HAMSTER
Mesocricetus newtoni
Head and body 15–18cm, tail 1–2cm. Like the Common hamster, but much smaller and with less black below. Habits similar to Common hamster. The closely related Golden hamster (*Mesocricetus auratus*) from the Middle East has given rise to the domesticated hamster used in research and in the pet trade. In these the colour and hair length are very variable, but they are usually paler with less-pronounced black patches.

GREY HAMSTER
Cricetulus migratorius
Head and body 9–10cm, tail 2–3cm. Small, and lacking the bold colour pattern of other hamsters, it resembles a very short-tailed vole with large eyes and ears.
Range and habitat: widespread in Asia, and in Europe found locally in the Ukraine, Bulgaria and Greece, mainly in dry grassland and cultivated areas up to the fringes of deserts.
Food and habits: eats young shoots, seed and grain. Makes chambered burrows with large food storage areas.
Breeding: up to 2 litters of 4–6 young each year.

Hamsters have large cheek pouches which open inside the mouth and, when filled with food, may double the size of the head.

Common Hamster

Grey Hamster

LEMMINGS AND VOLES
Sub-family Microtinae

Some 25 species of this sub-family are found in Europe. They are characterized by their shortish tail, short legs and blunt muzzles, and by the complex patterns of enamel and dentine on the molar teeth. They are the predominant small grazing animals of European grasslands. Their numbers tend to fluctuate, periodically reaching plague levels. When the population of lemmings peaks, the excess animals migrate in huge numbers, usually down valleys, and many may reach the sea.

NORWAY LEMMING
Lemmus lemmus
Head and body about 130mm, tail 20mm. The long fur is yellow-brown above, with a bold pattern of black streaks and patches; lighter below.
Range and habitat: tundra areas.
Food and habits: eats grass, roots, sedges, moss and lichens. Burrows just below surface in summer, below snow in winter.
Breeding: several litters of 4–8 young each year. May breed under snow cover in winter.

WOOD LEMMING
Myopus schisticolor
Head and body 85–95mm, tail 16–18mm. Similar to a short-tailed vole. Grey with a reddish tinge above, grey below.
Habitat: coniferous forest.
Food: mainly mosses, liverworts and lichens.
Breeding: only 1 in 4 animals is male. Seldom reaches high population levels.

BANK VOLE
Clethrionomys glareolus
Head and body 85–100mm, tail 40–65mm. Above, bright chestnut or reddish; below, yellowish, buff or grey-white. Patterns of enamel on cheek teeth are important for identification of this and allied forms.
Habitat: scrub, deciduous woodland and hedgerows.
Food and habits: eats a wide range of vegetable matter and some insects. Active day and night in short bursts.
Breeding: several litters of 3–6 young born each year.

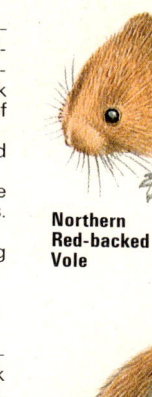

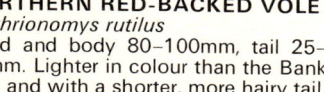

NORTHERN RED-BACKED VOLE
Clethrionomys rutilus
Head and body 80–100mm, tail 25–35mm. Lighter in colour than the Bank vole and with a shorter, more hairy tail.
Habitat: mainly areas of birch and willow scrub.

GREY-SIDED VOLE
Clethrionomys rufocanus
Head and body 100–120mm, tail 30–40mm. Reddish colour on back confined to narrow zone; flanks grey.
Habitat: shrub thickets and open woodland with shrub layer.

BALKAN SNOW VOLE
Dinaromys bogdanovi
Head and body 100–140mm, tail 75–110mm. Mostly grey with inconspicuous red tinge. Relatively long tail.
Range and habitat: found only in the Balkans on rocky slopes.
Breeding: small litters of 2–3 young are common.

Northern Red-backed Vole

Grey-sided Vole

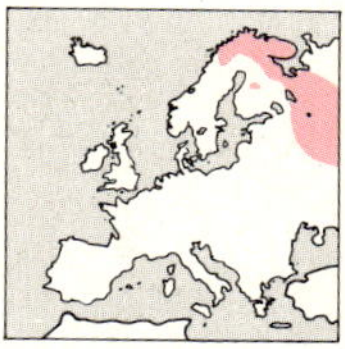

Northern Red-backed Vole

Grey-sided Vole

Balkan Snow Vole

FIELD VOLE
Microtus agrestis

Head and body 90–130mm, tail 30–45mm. Colour variable, yellow-brown to dark brown above, greyish-white below. Blunt head with short round hairy ears. Tail dark above, light below.

Range and habitat: an abundant and widespread species in grassland, meadows and marshland where ground cover is thick.

Food and habits: eats stems and leaves of grasses, reeds and sedges, and will take some animal food. May strip bark from trees in winter. Makes shallow burrows and a network of tunnels under vegetation on surface. Can cause extensive damage to crops and grassland pastures.

Breeding: several litters of 3–6 young born each year.

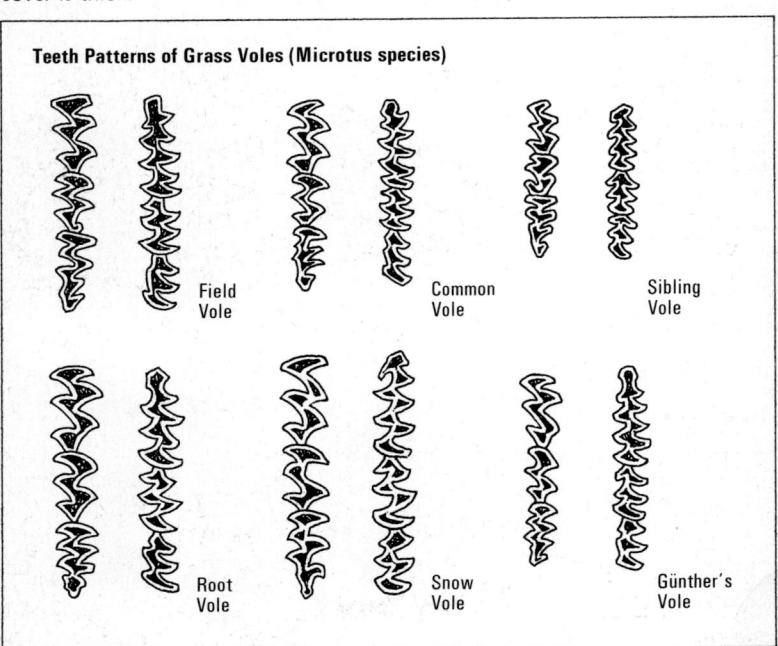

Teeth Patterns of Grass Voles (Microtus species)

Field Vole — Common Vole — Sibling Vole

Root Vole — Snow Vole — Günther's Vole

COMMON VOLE
Microtus arvalis
Head and body 90–120mm, tail 30–45mm. Very similar to the Field vole, but with shorter fur and less hairy ears. Colour lighter and tail less markedly bicoloured.
Habitat: grassland.
Habits: similar to Field vole, but burrows more and can survive in grasslands with less ground cover.

SIBLING VOLE
Microtus epiroticus
Can be distinguished from the Common vole only by differences in the chromosomes and by details of internal anatomy.

ROOT VOLE
Microtus oeconomus
Head and body 120–150mm, tail 40–60mm. Similar to the Field vole, but very slightly larger and with a distinctive enamel pattern on the first molar tooth.

SNOW VOLE
Microtus nivalis
Head and body 110–140mm, tail 50–75mm. Distinguished by its long pale tail, grey-brown colour and the pattern on the first molar tooth.
Habitat: lives mainly on mountains above the tree line and in woodland.
Food: alpine plants, grass, berries and shrubs.

GÜNTHER'S VOLE
Microtus guentheri
Head and body 100–120mm, tail 20–30mm. Similar to the Common vole, but with a particularly short tail. Tail and hind feet nearly white.
Habitat: dry grassland, meadows and crops, to which it can cause damage.

CABRERA'S VOLE
Microtus cabrerae
Head and body 107mm, tail 34mm. Large, with an area of very long, dark guard hairs on back.
Habitat: marshes and woodlands.

Field Vole

Common Vole

Root Vole

Snow Vole

Günther's Vole

COMMON PINE VOLE
Pitymys subterraneus
Head and body 80–100mm, tail 30–40mm. Similar to the Field vole, but with smaller ears and eyes. Hind foot short (less than 15mm) and with only 5 pads on the sole. Dark brown above, paler below.
Habitat: damp grassland, open woodland (but not common in coniferous forest).
Food and habits: eats roots, rhizomes and bulbs. Makes extensive burrows just below surface with nest and storage chambers. Mainly nocturnal.
Breeding: prolific, with up to 9 litters of 2–3 young each year.

OTHER PINE VOLES
A further 8 species of *Pitymys* are thought to occur in Europe, but the relationships between them have not been worked out. Identification is extremely difficult and is almost impossible from external characters. In many cases it depends upon a study of chromosomes and small details of skull and teeth. As the species can seldom be identified in the field, they are simply listed below with an indication of the distribution of each group currently recognized.

ALPINE PINE VOLE
Pitymys multiplex
Jura Mountains and southern Alps.

The Pine vole lives most of its life underground in extensive burrow systems and eats the underground parts of plants.

BAVARIAN PINE VOLE
Pitymys bavaricus
Bavarian Alps.

TATRA PINE VOLE
Pitymys tatricus
Tatra Mountains and nearby Poland.

LIECHTENSTEIN'S PINE VOLE
Pitymys liechtensteini
North-west Yugoslavia.

MEDITERRANEAN PINE VOLE
Pitymys duodecimostatus
South-eastern France, eastern and southern Spain.

LUSITANIAN PINE VOLE
Pitymys lusitanicus
Portugal, most of Spain and south-western France.

THOMAS'S PINE VOLE
Pitymys thomasi
Southern Yugoslavia and Greece.

SAVI'S PINE VOLE
Pitymys savii
Seems to exist in three separate populations: south-western France and northern Spain; Italy and Sicily; and southern Yugoslavia.

NORTHERN WATER VOLE
Arvicola terrestris
Head and body 16–19cm, tail 8–10cm. Like a large Field vole, but with a relatively long tail. Long, thick, glossy fur, blackish-grey or red-brown above, yellowish-grey below. Smaller and paler in southern part of range.
Habitat: edges of lakes and slow-moving rivers with good cover of vegetation on the banks, but can also be found in drier grassland.
Food and habits: eats mostly green vegetation, reeds, grass and some nuts and roots. Swims well, but is not pronouncedly adapted to aquatic habit. Burrows in river banks and makes a large grass-lined nest chamber. Some activity both day and night.
Breeding: several litters of 4–6 young born each summer.

SOUTHERN WATER VOLE
Arvicola sapidus
Head and body 17–20cm, tail 11–13cm. Almost identical to Northern Water vole, but where they are found together this is a larger, darker form with a relatively long tail, and is always found close to water.

MUSKRAT
Ondatra zibethicus
Head and body 30–40cm, tail 19–27cm. A North American species introduced to Europe by escapes from fur farms. Much larger than all native voles and with a tail which is flattened from side to side. Dark brown above, dirty white below. Lives close to water and swims well, but feet not webbed.
Habitat: found mainly near slow-moving streams, canals and lakes.
Food and habits: eats river-side vegetation, but will take roots and crops when available. Nocturnal. Burrows in banks and may cause considerable damage to waterways.
Breeding: several litters of 4–8 young born each year.

Northern Water Vole    **Southern Water Vole**

Cross Section Through Burrow of a Water Vole

chamber filled with grass

Water Vole

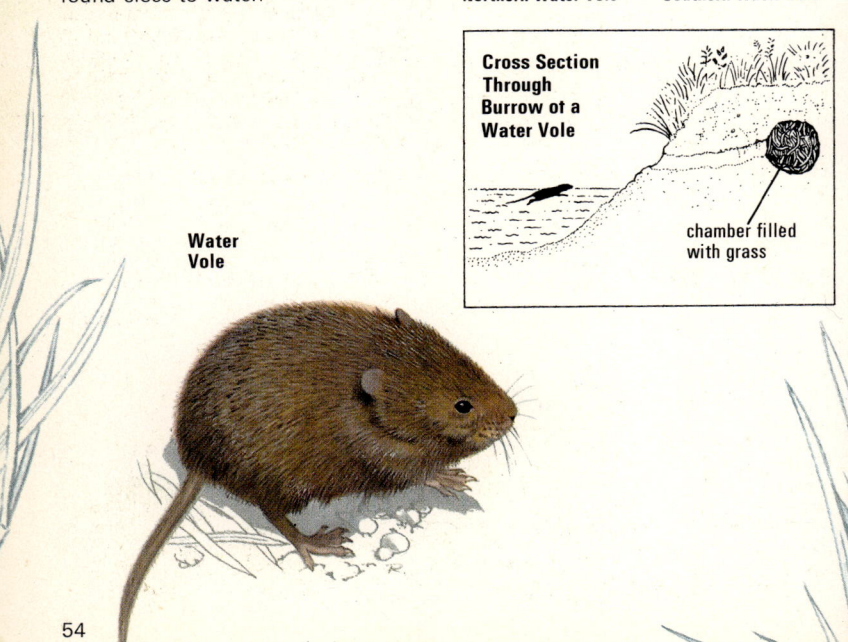

Muskrat

Muskrat

MOLE-RATS
Sub-family Spalacinae

A small sub-family of only 3 highly specialized rodents, 2 of which are found in Europe. They are adapted for burrowing and spend nearly all their life underground, although young animals may be found on the surface when dispersing to find new territories. They dig with their teeth, pushing loosened earth behind them with their feet. The jaw muscles are particularly strong, and the jaw is more mobile than in most rodent species. Mole-rats can be a pest to agriculture.

LESSER MOLE-RAT
Spalax leucodon
Head and body 18·5–27cm, tail vestigial. Streamlined, cylindrical body with soft velvety fur. Sandy above, darker below.
Habitat: grassland and well-cultivated soils.
Food and habits: feeds mainly on roots, bulbs and tubers and may pull whole plants down into burrow. Builds complex tunnel system with separate chambers for food, rest and latrine. Solitary, territorial and nocturnal.
Breeding: 1 litter of 4–5 is produced each year.

GREATER MOLE-RAT
Spalax microphthalmus
Head and body 24–30cm. Similar to Lesser mole-rat, but with greyer fur and clear-white markings on head.
Range and habitat: confined to grasslands of southern Russia.

Greater Mole-rat

RATS AND MICE
Sub-family Murinae

This is a large sub-family well represented in Asia and Africa. Most of the species are tropical or sub-tropical. The distribution of the genera *Mus* and *Rattus* is almost worldwide owing to their close association with man. In Europe 12 species occur. The members of this group are recognized by their long tails and large ears, and all have 3 molar teeth in each jaw. These adaptable animals occupy a wide range of habitats and have adopted many ways of life, including burrowing, climbing, swimming and partial dependence on man. Most feed on seeds, but other forms of plant material are included in their varied diet, and some, especially the commensal species, are omnivores. Many species are pests of agriculture and stored products, and are carriers of diseases.

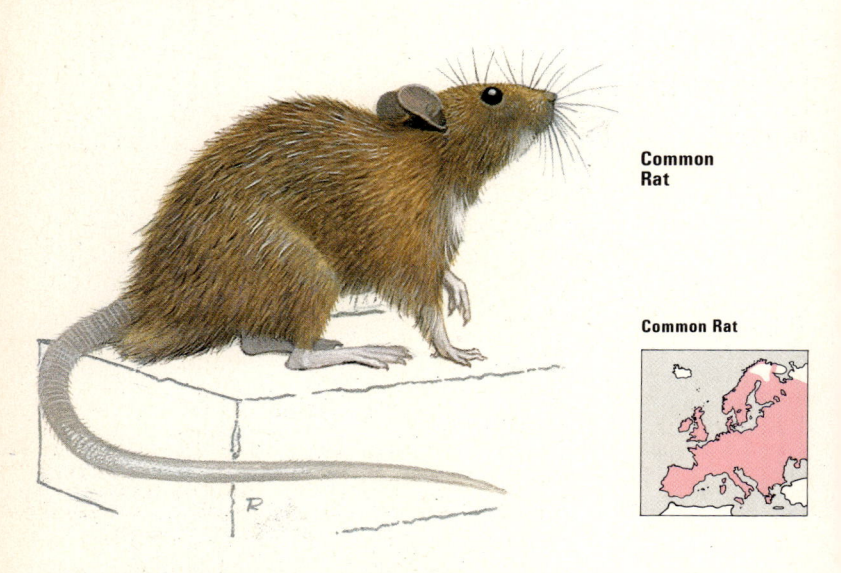

Common Rat

Common Rat

COMMON RAT or NORWAY RAT
Rattus norvegicus
Head and body 20–26cm, tail 17–23cm, hind foot 40–45mm. Usually brown, occasionally black above, grey below. Albino forms are used in research laboratories, and other colour forms have been bred for research and the pet trade.
Range and habitat: introduced to Europe from Asia in about AD 1500. Prefers habitats modified by man. Mainly terrestrial, but can swim well.
Food and habits: eats grain and seeds when available, but will eat almost anything. A devastating pest of stored produce, often polluting far more than it eats and causing a vast amount of damage. Implicated in the spread of many diseases, including plague. Mostly nocturnal.
Breeding: breeds all year round when food and shelter available, producing a litter of 7–8 young every 3–4 weeks.

BLACK RAT or SHIP RAT
Rattus rattus

Head and body 16–23cm, tail 18–25cm, hind foot 30–40mm. Distinguished from Common rat by its relatively long tail and short hind foot. Three colour forms are commonly found: dark grey all over, brown above with grey below, and brown above with white below.

Range and habitat: introduced from Asia at about the time of the Crusades, probably overland. More common in southern Europe than the Common rat. Closely associated with man and is the familiar rat of ships and seaports.

Food and habits: omnivorous. Climbs more than the Common rat. A considerable pest of stored produce and a carrier of many diseases.

Breeding: breeds throughout the year, producing about 5 litters of 5–10 young. Possibly declining in northern Europe due mainly to severe control measures.

Black Rat

brown and grey form

Black Rat

black form

WOOD MOUSE
Apodemus sylvaticus
Head and body 80–110mm, tail 70–110mm. Yellowish-brown above, silvery grey below. Usually has a small yellow patch on the chest which may extend to a streak. Eyes and ears large.
Habitat: very common in woodland, but frequently found in gardens and hedgerows, and often enters buildings.
Food and habits: eats mainly seeds, especially acorns, beech mast and hazel nuts, but also takes some insects and other invertebrate food. An agile climber. Nocturnal. A major source of food for owls and small mammals.
Breeding: breeds most of the year, producing several litters of 4–7 young.

YELLOW-NECKED FIELD MOUSE
Apodemus flavicollis
Head and body 90–120mm, tail 90–135mm. Slightly larger and more brightly coloured than the Wood mouse. White below, and with the yellow chest spot large and usually extended to form a collar.
Habitat: woodland, orchards and gardens, and often enters houses.
Habits: similar to the Wood mouse.

PYGMY FIELD MOUSE
Apodemus microps
Head and body 70–95mm, tail 65–95mm. Smaller than the Wood mouse. Yellow spot on chest inconspicuous or absent.
Habitat: usually found in open habitats, scrub, grassland and crops.

The Yellow-necked Field mouse will often be found in farm buildings, where it often makes stores of nuts. It is a pest of fruit crops.

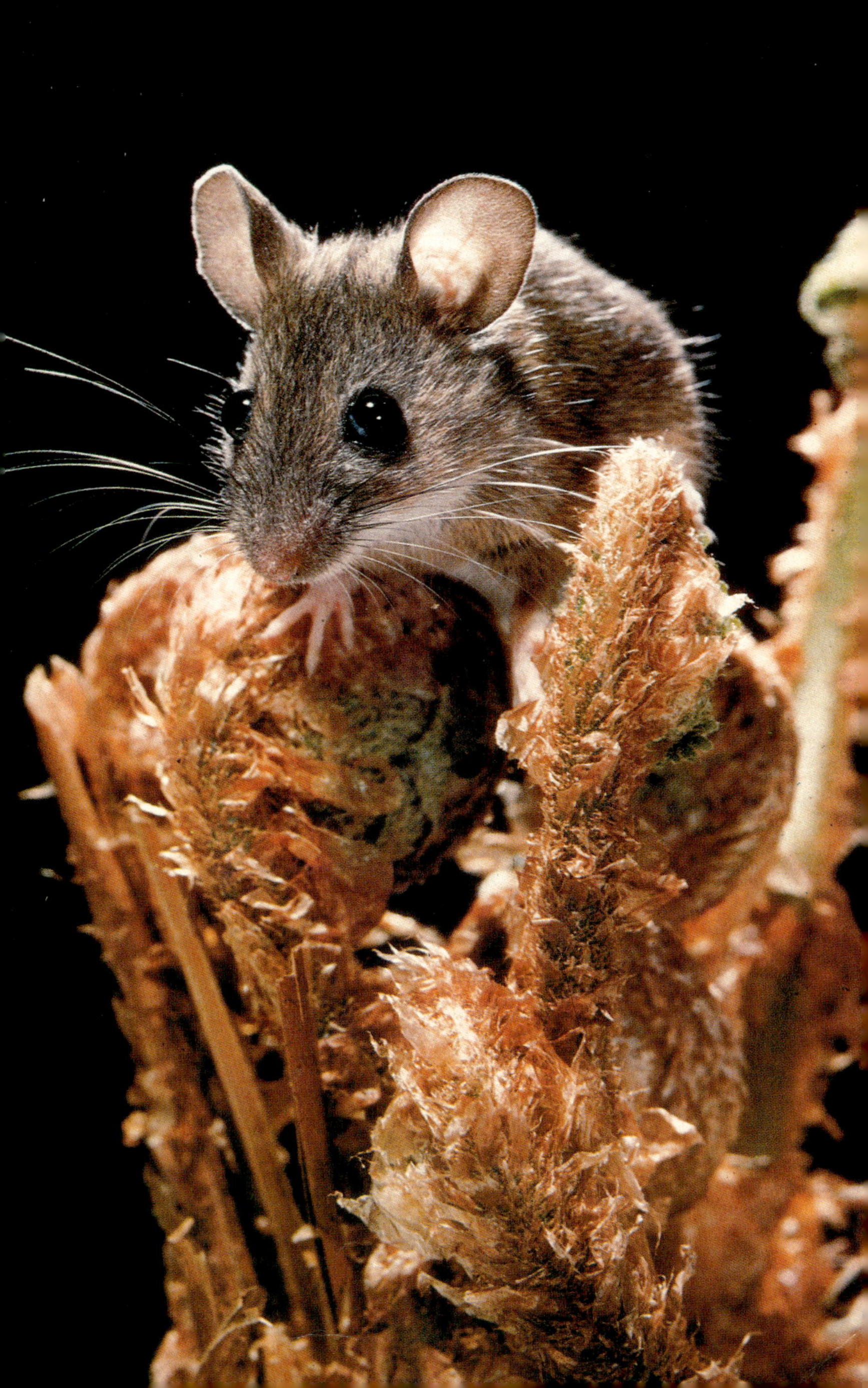

ROCK MOUSE
Apodemus mystacinus
Head and body 100–130mm, tail 105–140mm. Largest of the species of *Apodemus*. Sandy-grey above, white below. Lacks chest spot. Similar to a young rat, but more slender and with relatively small hind feet.
Range and habitat: Greece, Albania and the coastal mountains of Yugoslavia. Lives in open woodland and scrub, especially where rocks provide some shelter.
Habits: not well known.

STRIPED FIELD MOUSE
Apodemus agrarius
Head and body 90–115mm, tail 70–85mm. Very similar to the Wood mouse, but with a prominent dark stripe from nape of neck to base of tail. Tail shorter than head and body. Reddish-brown above, white below.
Habitat: lives in scrub, grassland, hedgerows and woodland margins, but not found in dense woodland. More terrestrial than Wood mouse.
Breeding: several litters of 5–7 young each year.

The Harvest mouse builds its breeding nest well above ground. Grasses are slit lengthways and woven into a ball up to 10cm in diameter. The female does the work in late pregananacy. In winter the nest is made underground at the end of a short burrow and is usually well stocked with stored seeds and grain. The numbers of Harvest mice are affected by agricultural practices such as combine harvesting, insecticide spraying, stubble burning and hedge clearance.

Harvest Mouse

Harvest Mouse

HARVEST MOUSE
Micromys minutus

Head and body 60-75mm, tail 50-70mm. A small rodent with thick, soft fur, brown above with a yellowish or russet tone; white below. Bicolored tail, partly prehensile, and used to assist climbing. Broad feet, small rounded ears.

Habitat: hedgerows, tall grass and reed beds. The mice remain in thick cover and are difficult to observe. Numbers vary greatly from place to place, but few localities have a high density. The destruction of their habitat by modern farming methods may be causing some reduction in numbers.

Food and habits: eats mainly seeds, fruits, and bulbs, but takes some insect food, particularly in summer. Stores seed for winter below ground. Active day and night, and does not hibernate.

Breeding: nest is a spherical ball of woven grass well above ground, in which several litters of 5-9 young are produced each year.

HOUSE MOUSE
Mus musculus
Head and body 70–95mm, tail 70–95mm. Size varies with habitat, and island forms often large. Soft brown-grey fur, slightly lighter below, but colour variable. Large eyes. Characteristic musty smell.
Range: origin probably central Asia, now worldwide associated with man.
Food and habits: eats almost anything, but prefers grain when available. A major pest of stored food and a carrier of several diseases. Mainly nocturnal.
Breeding: can breed throughout the year in suitable conditions, producing several litters of up to 10 young, but usually 5–6, and may increase to plague numbers.

ALGERIAN MOUSE
Mus spretus
Head and body 70–85mm, tail 55–70mm. A small mouse with a noticeably short tail. Yellowish-grey above, pale grey below.
Habitat: cultivated land, gardens and open woodland.

STEPPE MOUSE
Mus hortulanus
Head and body 70–90mm, tail 65–70mm. Similar to Algerian mouse, with a short tail and pale-coloured underside. But tail only slightly shorter than head and body.
Habitat: lives more frequently outdoors than House mouse in cultivated ground and grassland.
Habits: makes large grain stores and can cause great damage to crops.

CRETAN SPINY MOUSE
Acomys minous
Head and body 90–120mm, tail 90–120mm. Whole of upper surface is covered in spines. Pale reddish-brown above, white below. Large ears and thick tail.
Range and habitat: found only on the island of Crete, to which it was probably introduced from North Africa. Lives in dry scrub, but also enters houses.
Food: feeds on seeds.
Breeding: several litters born each year with 1–5 young, but usually 2–3.

BIRCH MICE
Family Zapodidae

A family of some 11 species distributed widely around the northern hemisphere in North America, Asia and Europe. Two species are found in Europe. Birch mice have an exceptionally long, prehensile tail and a distinct dark stripe along the back. The upper jaw has an extra grinding tooth. These are rare, seldom seen animals, active mostly at night and hibernating for a large proportion of the colder months.

Northern Birch Mouse

Southern Birch Mouse

Northern Birch Mouse

NORTHERN BIRCH MOUSE
Sicista betulina
Head and body 50-70mm, tail 75-100mm. Russet above, white below, but with a prominent dark stripe along the back from neck to base of tail.
Range and habitat: distributed patchily in Europe, but usually found in damp birch woodland with dense undergrowth.
Food and habits: feeds mainly on small invertebrates, but also eats seeds and bulbs. Nocturnal. Agile, and an especially good climber aided by the prehensile tail. Hibernates from October to April after becoming very fat in autumn.
Breeding: usually 1 litter each year with 3-5 young.

SOUTHERN BIRCH MOUSE
Sicista subtilis
Head and body 55-65mm, tail 65-80mm. Similar to the Northern birch mouse, but with a pair of pale yellowish stripes, one each side of the dark back stripe. Tail relatively slightly shorter.
Range and habitat: found mainly in the steppes of southern Russia, but scattered populations live in grassland, scrub and cultivated areas in other parts of Europe.
Habits: similar to Northern birch mouse.

CARNIVORES
Order Carnivora

About 250 species of carnivores are distributed widely around the world, although they were not present in Australia or New Zealand until introduced by man. About 25 species have been found in Europe. The range of size in the group is enormous, from the tiny weasel to the tiger. The main sub-groups in the order are: bears (Ursidae); dogs (Canidae), including wolves and foxes; mustelids (Mustelidae), including stoats, weasels and their allies; viverrids (Viverridae), including mongooses and genets; procyonids (Procyonidae), including raccoons; cats (Felidae); and hyenas (Hyaenidae). The name *Carnivora* means flesh-eater, and although the diet varies, most species are adapted for eating meat with strong jaws, sharp teeth and powerful jaw muscles.

Polar Bear

BEARS
Family Ursidae

There are about 10 species of bears distributed around the northern hemisphere and in northern South America. They are not found in Africa nor in Australia. Bears are large, heavily-built animals with small ears and eyes. They have 5 digits in each foot and strong claws. They walk with the feet flat on the ground. The canine teeth are large, but the molars are not noticeably modified for shearing flesh.

POLAR BEAR
Thalarctos maritimus
Head and body 1·5–2·5m, tail 8–10cm. A large bear, white with a tinge of yellow or cream. Has a pronounced 'Roman' nose. May weigh up to 650kg, but female is smaller.
Range and habitat: lives on coast and sea ice of the Arctic.
Food and habits: eats seals, fish and carrion. Solitary. Good swimmer.
Breeding: mates in July, and young born in February. There is usually only 1 cub, but up to 4 recorded born in den during period of winter inactivity.

BROWN BEAR
Ursus arctos
Head and body up to 2m, tail vestigial. Colour of shaggy coat varying shades of brown. Weighs up to 250kg.
Range and habitat: lives in wild country, often in mountains. Population in forests of western Europe heavily persecuted and the remaining specimens are smaller than their eastern counterparts.
Food and habits: a very varied diet, mostly vegetarian with roots, bulbs, tubers, berries and grain. Also eats small mammals, eggs, and fish. Can open bees' nests for the honey. Solitary and mostly nocturnal. Hibernates in winter in a cave or sheltered hole.
Breeding: young, often twins, born in January or February; small at birth.

Brown Bear

Brown Bear

Wolf

DOGS AND ALLIES
Family Canidae

The family, which includes dogs, wolves, jackals and allies, contains some 35 species widely distributed around the world, except in New Zealand and some Oceanic islands. The Australian dingo is thought to have been introduced with early man. There are 5 species in Europe. Most members of the family are long-legged, deep-chested animals adapted for hunting. The canine teeth are large and the shearing mechanism of the molars well developed. The claws are blunt and not retractable. The domestic dog is probably descended from the wolf. Many species have been hunted for fur and in order to protect man, his domestic animals and game.

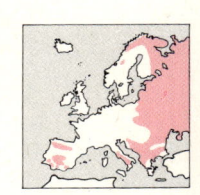

Wolf

Jackal

WOLF
Canis lupus
Head and body up to 140cm, tail 40cm. The size is variable, with animals in the south being smaller. A dog-like animal with broad head and chest; long, bushy, drooping tail; and pointed ears. Colour variable, but usually a brownish- or yellowish-grey brindled with black.
Range and habitat: the original European population is now much reduced and the animals are found in the more remote forests, mountains and tundra. The wolf was exterminated in Britain by the middle of the 18th century.
Food and habits: feeds on the larger ungulates (deer, reindeer and elk), but will also take smaller prey. In summer lives in small family groups in specific territory, but in winter several groups may join together to form a large pack for hunting. May be a serious pest of domestic animals.
Breeding: male and female co-operate to rear the young, usually 5–6 cubs born in spring.

JACKAL
Canis aureus
Head and body up to 75cm, tail up to 36cm. Smaller and more lightly built than the wolf, the jackal can also be recognized by the reddish-brown colouring on the side of the neck. Colour a dirty yellow mixed with black and brown hairs. Tail reddish with black tip.
Habitat: lives mainly in steppe and scrub vegetation and is less often found in forests.
Food and habits: feeds on smaller prey than the wolf, much of its diet consisting of small mammals, birds, eggs, fish and fruit. A pest of domestic poultry, and will also attack sheep and goats. Lives in pairs or small family groups in a specified territory marked with urine. More nocturnal than the wolf.
Breeding: usually 4–5 young born in spring and reared by both parents.

Jackal

Fox cubs weigh about 100g at birth and first leave the nest at about three weeks of age. They reach adult size at six months.

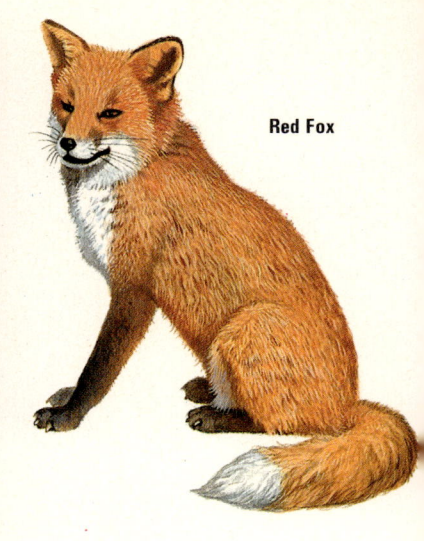

Red Fox

RED FOX
Vulpes vulpes
Head and body about 60 cm, tail about 40cm. Fur sandy to brownish-red above, greyish-white below. Has black markings on front of limbs and back of ears. Tail bushy, usually with a white tip. Coat thicker in winter. Muzzle pointed, and ears erect and sharply pointed.
Habitat: preferred habitat is woodland, but very adaptable and significant numbers are found in towns. Despite persecution, numbers have not noticeably been reduced.
Food and habits: mainly small mammals, squirrels and rabbits, but also eats insects, small birds, eggs, grass, fruit and carrion. Solitary most of year, but male assists in rearing young. Mostly nocturnal, but can hunt by day. Extensively hunted for sport and as pest of poultry and game farms. Known to carry rabies.
Breeding: usually 4 cubs born in den in spring.

ARCTIC FOX
Alopex lagopus
Head and body up to 67cm, tail to 42cm. Colour variable, but summer coat usually greyish-brown, turning white in winter. A small percentage of animals are a smoky-grey colour throughout the year and are known as the Blue fox. Muzzle and ears are shorter than in Red fox. Coat very thick, particularly in winter. Extensively hunted for valuable fur.
Range and habitat: lives in the tundra and near coast. The Blue fox is particularly common in Iceland.
Food and habits: feeds mainly on voles and lemmings, birds, carrion and shellfish. Said to be very tame.
Breeding: may have 2 litters each year with 5-8 young. Population numbers fluctuate widely depending on those of main prey species.

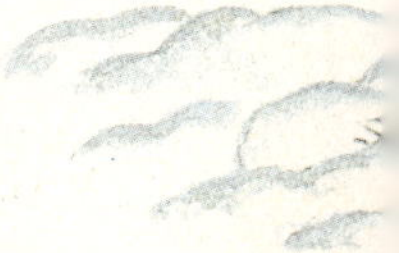

Arctic Fox

RACCOON-DOG
Nyctereutes procyonoides
Head and body 50–55cm, tail 13–18cm. General colour yellowish-brown, with hairs on shoulder, back and tail tipped with black. Large dark spot on each side of face. Limbs dark. Winter coat very long and thick.

Range and habitat: a native of eastern Siberia and Asia, the Raccoon-dog has been brought to Europe for fur-farming and has now spread as far west as Germany, Switzerland, and Sweden and Finland. Lives mainly in river valleys and grassy plains and forest.

Food and habits: varied diet includes rodents, fish and acorns. Eats a lot of berries and fruits in autumn. Spends worst of winter in hibernation, the only member of the family to do so.

Breeding: usually 1 litter of 6–8 young each year.

WEASELS AND ALLIES
Family Mustelidae

The family is widely distributed around the world, except for Australia, Madagascar and some Oceanic islands. It includes about 70 species of medium-sized, short-legged, sinuous, long-bodied animals such as weasels, polecats, badgers, otters, mink and their allies. Most species produce a strong-smelling odour from anal glands. They are meat-eaters with prominent canines and well-developed shearing teeth. Many species are hunted for their fur; some have been persecuted as pests; others have been farmed and domesticated.

Stoat

winter coat

Stoat

summer coat

STOAT
Mustela erminea

Head and body 22–32cm, tail 8–12cm. Males considerably larger than females. Body long and slim with short legs. Reddish-brown above, white tinged with yellow below. Tip of tail black. In northern part of range winter coat is white, but tail tip remains black. Fur is known as *ermine*. This is harvested mainly in Russia, but supplies are declining. Musky odour produced by anal glands.

Habitat: lives mainly in woodland, but adaptable and found in hedgerows and moorland.

Food and habits: diet almost entirely meat, especially rodents, rabbits, fish, birds and eggs. Mainly nocturnal, but may hunt by day.

Breeding: 1 litter per year with 4–5 young. Population numbers fluctuate with abundance of prey species. Regarded as pest of poultry and game birds, and much persecuted.

WEASEL
Mustela nivalis

Head and body about 20cm, tail 6cm. Female smaller than male. The smallest European carnivore. Weasels of southern Europe are larger than those from the north. Smaller than the stoat, with relatively short tail with no black on tip. Reddish-brown above, white below, often with brownish markings. May turn white in north of range (Russia and Scandinavia). Musky smell from anal glands.

Habitat: found in nearly all habitats, even in towns. Active day and night, preying on mice and voles, other small rodents, birds and eggs. May cause considerable damage to poultry farms and to game birds.

Breeding: 1 or 2 litters of 4–6 young each year. As with stoat, numbers fluctuate according to population of prey, especially voles.

EUROPEAN MINK
Mustela lutreola
Head and body 35–40cm, tail 13–14cm. Male larger than female. A large weasel-like species. Dark brown all over, except for white markings above and below the mouth. No white around eye. Feet partially webbed; swims and dives well.
Habitat: lives in marsh lands and on the banks of lakes and rivers.
Food and habits: eats water voles, rodents, birds, frogs and fish. Nocturnal and solitary. Sleeps by day in a hole or burrow or in a hollow tree.
Breeding: usually 1 litter of 4–5 young each year.

AMERICAN MINK
Mustela vison
Head and body 35–40cm, tail 13–14cm. Very similar to the European mink, but white on face restricted to the lower jaw. Many colour varieties bred in captivity for fur (*ranch mink*).
Range and habitat: native to Canada and the United States; has escaped on many occasions from fur farms in Europe and is now widespread in the wild.
Food and habits: eats a similar wide range of prey to the European species; habits also similar. Impact on ecology of wet areas not yet understood, but may compete with otters in some regions.

WESTERN POLECAT
Mustela putorius
Head and body 30–44cm, tail 13–18cm. Coarse, dark brown hair above with yellowish underfur showing through. Black below. White on muzzle and between eyes and ears.
Habitat: lives mainly in woodland areas and near water.
Food: rodents, rabbits, frogs, eggs, birds and invertebrates.
Breeding: 1 litter each year with 5–6 young. Often killed as pest of game and poultry and formerly for its fur.

STEPPE POLECAT
Mustela eversmanni
Head and body 30–44cm, tail 13–18cm. Very similar to the Western polecat, but paler in colour above.
Habitat: lives mainly in grassland, scrub and cultivated land.
Food and habits: grassland animals such as susliks and hamsters are prominent in diet. More active in daylight hours than Western species. Habits otherwise similar.

European Mink

European Mink

Western Polecat

Western Polecat

Steppe Polecat

Steppe Polecat

Marbled Polecat

DOMESTIC FERRET
Mustela furo
Head and body and tail within size range of polecats. Origin of domestic form unclear, as it interbreeds with Western polecat in wild, but has similar skull features to the Steppe polecat. Colour very variable, and in wild populations tends to resemble polecat. Domestic form usually albino. Kept for hunting and control of rabbits.

MARBLED POLECAT
Vormela peregusna
Head and body 30–38cm, tail 15–20cm. Similar to the other polecats in shape, but with distinctly mottled coat of dark brown and cream. Dark brown below.
Habitat: lives in open areas with scrub and trees, and is also found in farmland.
Food: eats small mammals, birds, lizards and reptiles and some invertebrates.

PINE MARTEN
Martes martes

Head and body 40–55cm, tail 22–27cm. Colour varies from brown to nearly black above and below. Conspicuous throat patch yellowish to orange, but never pure white. Sharp muzzle, prominent ears, long bushy tail.

Habitat: lives in woodlands, especially the pine woods of northern Europe; being a good climber, is often seen in trees.

Food and habits: feeds on small birds, squirrels and other small rodents, rabbits and eggs, but will also take honey and berries. Mainly nocturnal, hunting in small groups or alone.

Breeding: usually 1 litter of 3 young born each year in spring. Trapped intensively for fur, and has become rare in many parts of range; also trapped to protect game birds.

Pine Marten

Beech Marten

Pine Marten

Beech Marten

Cretan form

BEECH MARTEN
Martes foina
Head and body 40–48cm, tail 22–26cm. Similar to Pine marten, but with a pure-white throat patch, which is often divided by a dark streak. In Crete the throat patch is much reduced and may be absent.
Habitat: lives in deciduous woodland and rocky areas, and is often found near houses. Less often seen in trees than Pine marten.
Food and habits: preys on mice, shrews, birds, frogs and lizards; also eats some fruit and berries. Habits similar to those of Pine marten, but will make den in houses and barns and readily enters towns.
Breeding: usually 1 litter each year with 3–6 young.

WOLVERINE
Gulo gulo
Head and body 70–80cm, tail 16–25cm. Often known as the Glutton. Largest of the weasel family. Dense, shaggy coat of thick fur, dark brown above, with band of paler colour on side and dark below. Thick body with short, strong legs.
Range and habitat: lives in evergreen forests in colder parts of Europe and on the tundra.
Food and habits: eats mostly rodents, birds, eggs, and invertebrates in summer, but mainly carrion in winter. Can attack sickly animals of larger size and may eat deer. Its particularly strong jaws are used to tear carrion. Hunts alone or in pairs. Nocturnal and relatively slow-moving.
Breeding: a single litter of 2–3 young is born in spring each year. Much persecuted by man for fur and as pest of domestic animals, but is still common in many parts of its range.

Wolverine

Wolverine

OTTER
Lutra lutra

Head and body 60-80cm, tail 35-45cm. Fur rich brown above, paler below, especially on throat. Thick underfur grey, mixed with long, glossy guard hairs. Long body with long tapering tail, short legs, webbed feet. Ears very small.

Habitat: lives by lakes and rivers.

Food and habits: feeds mainly on fish of many kinds, although eels are often preferred. Also eats crayfish and other invertebrates, and water birds. On land will take voles, rabbits and other small mammals. Usually nocturnal.

Breeding: has no distinct breeding season: young can be born at any time of year. Usually 2-3 cubs in each litter. Hunted for fur, sport and to protect fishing interests, the otter is becoming scarce in some parts of its range. Habitat destruction is also an important factor in its decrease, as is water pollution.

The otter will catch any available species of fish and carry them to the bank to eat. In coastal areas it hunts flatfish, crabs and shellfish in shallow waters.

Otter

BADGER
Meles meles

Head and body 70–80cm, tail 12–19cm. Stout, squat body; head with long tapered snout. The rough coat is grey above, and black below and on limbs. Head white, with conspicuous black stripe on each side running through the eye to the small ears, which are tipped with white.

Habitat: lives mainly in woods and copses in extensive burrow systems called *sets*. May also live in fields and hedgerows.

Food and habits: feeds on a wide range of animals and plants, including earthworms, small mammals, carrion, fruit, nuts and bulbs. Lives in family groups. Nocturnal.

Breeding: a single litter of 1–4 young born in late winter or early spring. Main enemy is man. Some people believe the badger is implicated in the spread of bovine tuberculosis in some areas of Britain. It is also killed for sport and allegedly to protect game birds and poultry, but the badger's effect on these is probably insignificant.

Genet

MONGOOSES AND GENETS
Family Viverridae

A family of small- to medium-sized carnivores, including about 75 species of civets, genets, and mongooses widely distributed in Asia, Africa and southern Europe. The family includes a number of species with striped and spotted coats, sinuous bodies and long bush tails. Scent-gland secretions, known as *civet*, are taken from several species and used in the perfume and pharmaceutical industries.

EGYPTIAN MONGOOSE
Herpestes ichneumon
Head and body 50–55cm, tail 35–45cm. Long sinuous body, grizzled grey all over. Tail tipped with black, tapering to a point.
Range and habitat: found in southern Spain and Portugal, probably as an introduction from Africa, where it is common from Egypt to the Cape. Lives in scrub and open woodland.
Food and habits: feeds on rodents, birds, eggs and reptiles. Known for ability to kill snakes by agility and speed of attack. Nocturnal, but sometimes active by day. Usually solitary, but sometimes found in small family groups.
Breeding: 1 litter each year usually born in late summer.

Egyptian Mongoose

Egyptian Mongoose

INDIAN GREY MONGOOSE
Herpestes edwardsi
Head and body about 45cm, tail 45cm. Slightly smaller than Egyptian mongoose. Grizzled grey-brown, with tip of tail lighter.
Range: introduced in about 1960 to Italy from India or south-west Asia.
Habits: similar to the Egyptian mongoose.

GENET
Genetta genetta
Head and body 50–60cm, tail 40–50cm. Similar to a small, lightly built cat. Coat spotted dark brown on pale ground colour; tail banded light and dark. Short legs, large ears and pointed muzzle.
Range and habitat: range includes most of Africa and Arabia. Lives in bush and scrub country.
Food and habits: eats mainly small mammals, birds and insects. Nocturnal, usually solitary.
Breeding: normally 1–3 young in 2 litters each year, born in spring and late summer.

RACCOONS
Family Procyonidae

A small family of about 18 species, including raccoons and kinkajous. Some scientists include the Lesser and Giant pandas of Asia in this family. Only one species is found in Europe, and that is a native of tropical and temperate parts of America. Most species are small- to medium-sized carnivores with long tails and well-developed carnassial teeth.

RACCOON
Procyon lotor
Head and body 50–60cm, tail 30–40cm. About the size of a cat, but easily identified by its black mask contrasting with white face markings, and by its distinctly banded tail. General colour greyish-brown.
Range and habitat: widespread in the Americas, and was introduced to Germany and neighbouring countries. It has since spread to the Netherlands and Luxembourg. Lives mainly in woodland, often close to water.
Food and habits: feeds mainly on aquatic animals, especially crayfish, fish and insects in the water, and rodents and a little plant food on land. Nocturnal.
Breeding: usually 1 litter of 3–4 young born each year in spring.

Raccoon

Raccoon

CATS
Family Felidae

A family of some 36 species of cats widely distributed around the world. Although differing greatly in size and coat pattern, all are unmistakably cat-like in appearance, with supple, muscular bodies, short round heads, eyes with pupils which contract vertically, and well-developed whiskers. The claws can be retracted. Only 2 species are native to Europe.

LYNX
Felis lynx
Head and body 80–130cm, tail 11–25cm. A medium-sized cat with long legs, markedly tufted ears and a ruff round the face. Tail short, black towards the tip. Coat reddish with dark spots which vary in number and are often indistinct, especially in the north of the range. Some scientists consider the Spanish lynx to be a separate species, as it is smaller and more heavily spotted.
Range and habitat: found in North America and Asia, and formerly widespread in Europe. Now very scarce in most of range and protected in many areas. Lives mainly in coniferous forest and mountainous country, but also found in scrub.
Food and habits: feeds on hares, rodents and young deer. Nocturnal.
Breeding: usually 2–3 kittens are born in spring each year.

WILD CAT
Felis sylvestris
Head and body 50–65cm, tail 30cm. About the size of a large domestic cat, but somewhat more robust. Can be distinguished from domestic cats by its short bushy tail, which is marked with clear, separate dark rings; the coat, which is usually clearly striped; and the pale paws. The tail has a blunt tip.
Habitat: lives in dense woodland and on rocky hills.
Food and habits: eats mostly small mammals, hares and rabbits, but may also kill small deer and lambs, birds, fish and some insects. Nocturnal and solitary.
Breeding: usually 2–4 kittens born in late spring.

The Wild cat is often killed as vermin by gamekeepers, although it eats mostly small mammals.

DOMESTIC CAT
Felis catus
Descended from Wild cat ancestors, probably first in North Africa, and will interbreed with Wild cat. Specimens which are similar to Wild cats usually have longer, more pointed tails and round blotches of colour rather than stripes. Rings on tail usually less well marked.
Range: found wild throughout Europe.
Breeding: often more than 1 litter per year, and hybrids between Wild cat and domestic stock may also have more than 1 litter. Many varieties produced by selective breeding.

Lynx

Wild Cat

Wild Cat

SEALS, WALRUSES AND SEA LIONS
Order Pinnipedia

An order containing about 30 species of aquatic carnivores living in coastal waters throughout the world, but more commonly in temperate or cold waters. They have streamlined bodies, and all the limbs are modified into flippers and webbed. The ears are reduced or absent, and the nostrils slit-like.

SEALS
Family Phocidae

The family includes 18 species of true, or earless, seals spread widely around the world. The fore flippers are smaller than the hind, and the hind limbs cannot be turned forwards. The adult fur is stiff and lacks underfur. Seals come ashore at all times of the year, and males establish breeding territories in summer. Formerly the basis of a widespread industry, but many species became endangered and are now protected.

COMMON SEAL
Phoca vitulina
Length 2m, females smaller. Colour variable, grey-yellow to dark brown, usually marked with dark spots. Pups similar in colour to adults. Nostrils arranged in V-shape, and muzzle concave in profile.
Food and habits: feeds mainly on fish, molluscs and crustaceans. Can dive for 5–10 minutes or more.
Breeding: breeds in mid-summer; a single pup is born ashore, but can swim from birth.

RINGED SEAL
Phoca hispida
Length 1·4m. The smallest seal in European waters. Light grey in colour, with black spots ringed with contrasting pale colour. Pups white.
Food: mainly crustaceans.
Breeding: breeds on land-fast ice in spring. Pups remain on ice for about 2 months, and moult to a dark coat.

pup

Grey Seal

head profile

Ringed Seal

Grey Seal

GREY SEAL
Halichoerus grypus
Length 3·8m, females very much smaller. Muzzle long and with pronounced 'Roman' nose. Nostrils widely separated. Colour dark brown or black with lighter spots or blotches. Females paler. Pups white, but moult to darker coat after 2–3 weeks.
Habitat: often hauls out of water on to rocky shores and sand banks.
Food: feeds mainly on fish and squid.
Breeding: Grey seals are gregarious animals which breed in large colonies. The pups are born on shore in autumn, but deserted after 2 or 3 weeks.

Ringed Seal

MONK SEAL
Monachus monachus
Length 2·7m. Colour is a uniform brown on back, greyer below and usually with a white patch on belly. Pups black.
Range: Mediterranean and Black Sea.
Breeding: forms small breeding colonies on secluded beaches in autumn. Pups remain on beach for about 6 weeks. Much persecuted by fishermen and frequently disturbed on breeding beaches by holidaymakers, the Monk seal is now very rare and the population is still declining. The total number of remaining seals is estimated to be about 1,000–2,000. Related species in the Caribbean are similarly threatened.

HARP SEAL
Pagophilus groenlandicus
Length 2m. A light tawny grey with distinct black markings in male. Female has less well-defined marks which gradually darken during life. Pups white.
Habitat: lives in open sea and pack ice.
Food: feeds mainly on fish, squid and planctonic animals.
Breeding: forms large colonies on pack ice in spring to breed. Pups are fed for 2 weeks then deserted. They enter the water about 2 weeks later having lost white coat and grown a grey one.

BEARDED SEAL
Erignathus barbatus
Length 3m. A large seal with prominent whiskers. Colour is a uniform grey with brownish tinge without any spots or markings. Pups greyish-brown at first, but then moult to slightly spotted coat before attaining uniform colour.
Habitat: a rare seal, living in shallow coastal waters.
Food: feeds on bottom of sea, eating mainly molluscs.
Breeding: does not form large colonies, but breeds in small groups on ice floes in spring. Pups remain with mother for some months.

HOODED SEAL
Cystophora cristata
Length 3m, female smaller. Enlarged nasal cavity in male forms the 'hood', which can be blown up to form a football-sized bladder. Smaller hood in female. When not inflated the sac hangs down over nose. General colour grey with black patches. Pups are silvery grey above, white below, with a dark head.
Habitat: lives amongst drifting ice and in open sea. Solitary or in small groups.
Breeding: pups are born on ice and deserted after 2 weeks. They enter water 2 weeks later.

Bearded Seal

Monk Seal    Harp Seal

male, with hood inflated

female   Hooded Seal

pup

WALRUSES
Family Odobenidae

A distinctive family with only one species. Walruses have thick, swollen bodies covered in thick, wrinkled skin, with a layer of blubber beneath it. The upper canine teeth form large tusks in both sexes. In males they may reach up to a metre in length. The hind feet can turn forward to aid movement on land. There is no external ear.

Walrus

Walrus

WALRUS
Odobenus rosmarus
Head and body up to 4m in males, 3m in females. Blunt face with prominent upper lip covered in stout bristles. Tusks of both sexes are characteristic. Fur short, and colour a uniform grey-brown; skin very wrinkled.
Range and habitat: lives in shallow coastal seas of the Arctic.
Food and habits: feeds mainly on molluscs; occasionally kills seals and will also eat carrion. Gregarious, forms herds of 100 or more on rocky coasts and on ice.

Breeding: breeds in spring, on ice. The pups remain with the mother for a year or more. Hunted in the past for food, skin and tusks, and is still hunted for tusks. The total population is thought to be declining.

Large groups of male walruses without mates gather together on shore to sleep off their large meals of clams and fish. The tusks of an adult male walrus may each weigh 5kg and grow throughout the life of the animal.

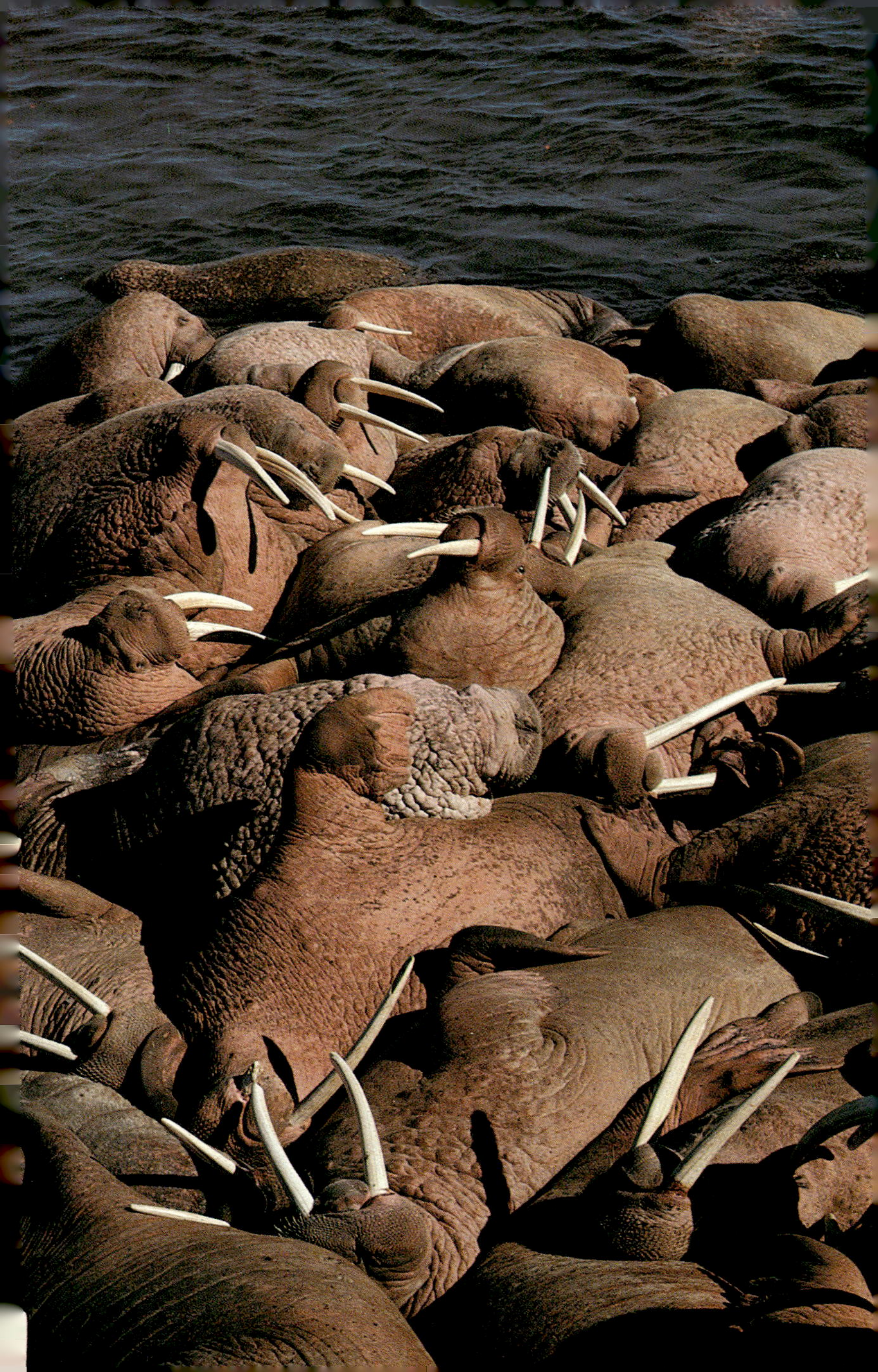

ODD-TOED HOOVED MAMMALS
Order Perissodactyla

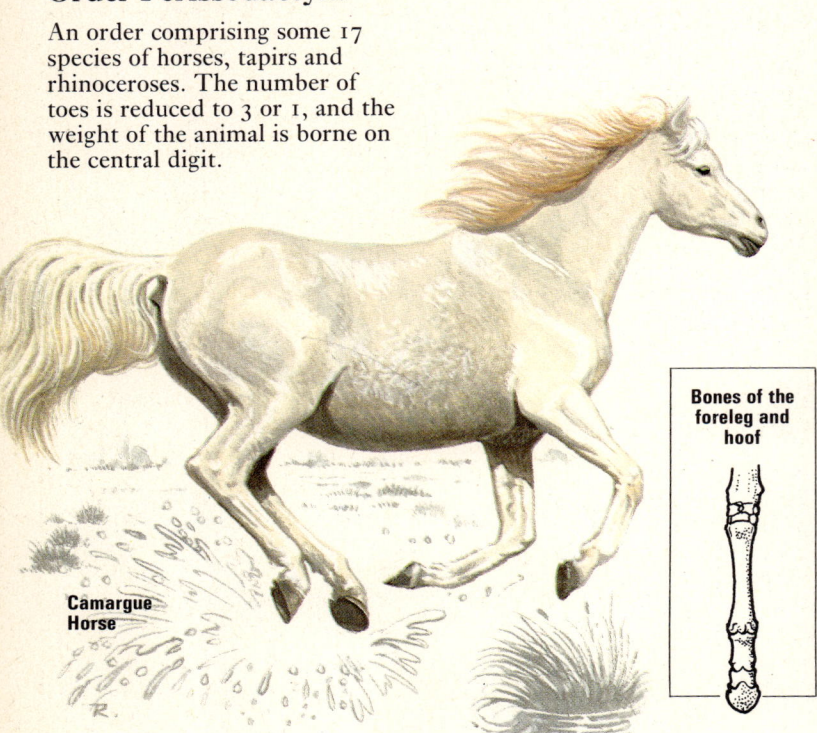

An order comprising some 17 species of horses, tapirs and rhinoceroses. The number of toes is reduced to 3 or 1, and the weight of the animal is borne on the central digit.

Camargue Horse

Bones of the foreleg and hoof

HORSES
Family Equidae

A family of Asian and African origin, with characteristic long necks, long heads, a single functional digit on each foot, and a long tail. Horses are grassland animals.

HORSE
Equus caballus
Size varies with breed. Horses are distributed worldwide as a result of introduction by man. No truly wild horses are found in Europe, although in several areas relatively unimproved breeds survive, some in an almost wild state. The New Forest ponies and Exmoor ponies in Britain and the Camargue horses in France are examples. The Przewalski horse of China and Mongolia most closely resembles truly wild stock and many herds are kept in captivity.

ASS
Equus asinus
A domestic animal derived from the wild ass of northern Africa. Smaller than most horse breeds, with very long ears, an erect mane and short hairs at base of tail. Commonly used as beast of burden.

EVEN-TOED HOOVED MAMMALS
Order Artiodactyla

An order which includes all cloven-hooved animals, such as pigs, sheep, deer, antelopes, giraffes and cattle. Most of the world's large herbivores belong to the group. The body weight is borne by the 2 middle digits. Many species (*ruminants*) are cud-chewing and have a complex stomach structure adapted to digesting large quantities of vegetable matter.

PIGS
Family Suidae

Nine species of pigs are found in Europe, Africa and Asia. They prefer a forest habitat with plenty of cover. All wild species have long snouts and many teeth. The tusks are developed from canine teeth.

WILD BOAR
Sus scrofa

Head and body 1·8m, tail 30cm. Size very variable, larger in eastern part of range. Has a dense, bristly coat of grey to black colour and a long, mobile snout. Canine teeth well developed, giving tusks up to 30cm long in large males. Young has yellow stripes along the body. Ancestor of the domestic pig.
Habitat: lives mainly in deciduous woodland.
Food: diet mostly vegetable (roots, bulbs, acorns and beech mast), but also takes animal food. Can be pest by digging up root crops.
Breeding: produces large litters of up to 12 piglets in spring. Hunted for sport, and killed to protect crops.

Wild Boar

Bones of the foot and hoof

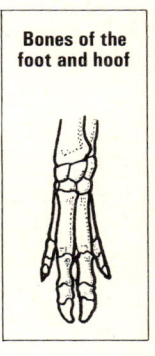

Wild Boar

young

CATTLE, SHEEP, GOATS AND ALLIES
Family Bovidae

More than 100 species are in this family, which is nearly worldwide. A great variety of forms are included, from graceful antelopes to heavy buffalo. They are grazers and browsers with complex stomachs. Most species have unbranched horns with a bony core permanently fixed to the bones of the skull and covered with a sheath of horny material. Horns are found on both the male and female in most species. Many species are gregarious and form large herds, this being their main defence against predators. Domestic species are nearly all descended from Eurasian species. A group of great economic importance, providing man with meat and leather. Many species are also hunted for sport.

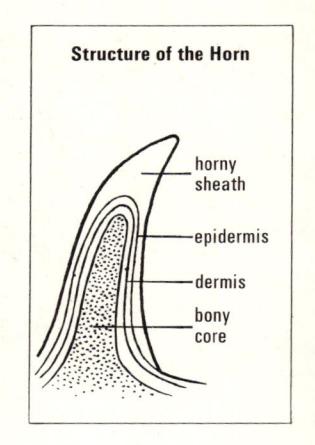

Structure of the Horn
- horny sheath
- epidermis
- dermis
- bony core

Bison

Musk Ox

BISON
Bison bonasus
Shoulder height 1·8m, female smaller. Covered in a shaggy coat of long brown hair. Males have a distinctive mane of long woolly hair. Horns present in both sexes.
Range and habitat: once widespread in deciduous forests of Europe; became extinct in wild, but survived in captivity and is now being re-introduced to forests of Poland, Rumania and Russia.
Food: browses on leaves of deciduous trees, especially oak and willow, and eats acorns and shoots of shrubs.
Breeding: a single calf is born in summer and reaches maturity in about 6 years.

DOMESTIC CATTLE
Bos taurus
Size variable. Descended from the auroch, which has been extinct in the wild since 17th century. No wild populations in Europe, but several herds of primitive domestic cattle survive. They graze rather than browse, and consume about 70kg of grass each day. They produce milk, meat and hide, and have been used as draught animals.

DOMESTIC WATER BUFFALO
Bubalus bubalus
Shoulder height 1·7m. An Asian species domesticated in many parts of the world and used in parts of southern Europe. Dark grey in colour all over, with long coarse hair. The large, flattened and wrinkled horns are backward-curving and heavy at the base, and may spread to over 1m. Produces milk and leather, but the meat is considered relatively poor. Used as a draught animal.

MUSK OX
Ovibos moschatus
Shoulder height 1·5m. Has a long shaggy coat of coarse hair over an inner coat of fine, dense fur. Black in front and below, legs lighter. Horns broad and down-curved.
Range and habitat: formerly widespread in Europe before last Ice Age, and has been re-introduced from American herds to parts of Norway and Sweden. Lives in the tundra habitat and can survive extreme cold.
Food: eats grass, sedges, and small shrubs.
Breeding: a single calf is born in spring.

MOUFLON
Ovis musimon
Head and body 120cm, tail 6cm, shoulder height 70cm. A truly wild sheep with a short rough coat, reddish-brown in colour, with dark markings, a white patch on each flank, and white legs and muzzle.
Range and habitat: lives in open woodland in mainland Europe, in open country on Mediterranean islands.
Food and habits: eats mainly grasses, herbs and sedges, and some shrubs. Gregarious, and similar in habits to primitive domestic breeds of sheep.
Breeding: 1 or 2 lambs born in spring.

male

female

Mouflon

Mouflon

DOMESTIC SHEEP
Ovis aries
Size variable. Coat entirely woolly, without the long straight hair of wild relatives. Domesticated since at least 5000 BC, probably first in south-west Asia. Intensive breeding and selection in last 200 years. A few primitive breeds survive, notably the Soay sheep on the island of Soay off Scotland.

ALPINE IBEX
Capra ibex
Head and body 140cm, tail 15cm, shoulder height 75cm. Females distinctly smaller. A wild goat with a mostly grey coat with some darker colour and a black streak along lower flanks. White below. Horns heavy and curved with prominent ridges on the fronts. In females, horns sweep straight back and are smaller.
Range and habitat: lives in Alpine meadows and hillsides above treeline.
Food and habits: mainly shrubs, grasses and sedges; in winter may eat lichens.
Breeding: single kid born in early summer. Became very rare due to hunting; now protected and re-introduced to Alps.

Alpine Ibex

SPANISH IBEX
Capra pyrenaica
Head and body 120–140cm, tail 12cm, shoulder height up to 75cm. Very similar to the Alpine ibex, but the horns diverge towards tips and curve up and out.

Range and habitat: lives in isolated populations in mountains of northern and central Spain.

Food and habits: similar to Alpine ibex.

Breeding: similar to Alpine ibex.

Alpine Ibex **Spanish Ibex**

Spanish Ibex

Feral Goat

WILD GOAT
Capra aegagrus
Head and body 130cm, tail 15cm, shoulder height 60-70cm. Truly wild goats are probably extinct, but some unimproved domesticated animals are feral on Crete and are thought to resemble closely the original wild stock. Distinguished from domestic goats and ibex by the shape of the horns, which sweep back in a smooth curve, diverge little and have widely spaced, rather inconspicuous ridges. Males have beard, dark stripe down back and across shoulder; females have duller markings and smaller horns.
Habitat: lives on rocky hillsides and in scrub.
Food and habits: browses on shrubs and trees, and grazes. Forms loosely organized flocks with males on higher ground.
Breeding: usually 1 kid, sometimes twins.

DOMESTIC GOAT
Capra hircus
Differs from the Wild goat in the more complex curvature of the horns, which turn outwards and up, diverging widely at the tips. Females have short horns. Feral goats usually have long coats of variable colour.
Range: some feral populations live in Britain and on many Mediterranean islands.
Food and habits: similar to Wild goat.

CHAMOIS
Rupicapra rupicapra
Head and body 90-130cm, tail 3-4cm, shoulder height 75-80cm. Goat-like, with stiff coarse hair, which is tawny brown in summer, changing to a longer, nearly black coat in winter. Lighter in colour below, with white throat patch. Horns in both sexes close together at the base, rising straight to a sudden bend backwards resembling a hook. Horns up to 20cm long.
Habitat: lives on wooded slopes and above the tree line in mountainous areas. Moves down to lower slopes in winter.
Food and habits: mainly herbs in summer, lichens and mosses in winter, but will eat young tree shoots. Gregarious, and forms large herds in winter. Males more solitary, except in rutting season. Renowned for ability to perform prodigious leaps. Its soft hide is used for glass-polishing leather, and its flesh is good to eat.
Breeding: usually a single young, born in spring.

male, summer

Chamois

Chamois

male, winter

95

DEER
Family Cervidae

A family of about 53 species widely distributed in the Americas, north-west Africa, Eurasia, Japan, Indonesia, Philippines, and introduced into Australia, New Zealand and other areas. Ten species are found in Europe, of which half are introduced. Nearly all species have antlers, at least in males. These are appendages of the skull, with a solid bony core supported on skin-covered *pedicels*. They are developed each year and shed after the *rut*, or breeding season. Growing antlers are covered in blood-rich skin called *velvet*. Each foot has 4 digits, of which the second and third are well developed.

RED DEER
Cervus elaphus

Shoulder height up to 140cm. A large species with a reddish-brown summer coat changing to a thicker brown-grey coat in winter. Whitish below; patch around tail buff-coloured. Calves usually brown, flecked with white. Only males have antlers, which are fully formed in August. A few rare animals (*hummels*) have poor or no antlers. The antlers have up to 12 points, occasionally more.

Habitat: lives in dense deciduous forest, but in some areas has adapted to moorland and open woodland.

Food and habits: mainly browsing animals, eating young shoots and leaves of deciduous trees and shrubs, but will also eat nuts and fruits. Females and young form separate herds; males solitary or in loose herds, except at rut. Much prized for meat, and main predator is man.

Breeding: usually a single calf born in early summer.

Red Deer

male

Red Deer

calf

THE DEVELOPMENT OF DEER ANTLERS EACH YEAR
The four stages cover the period spring–autumn

1

2

3

4

Structure of the Antler

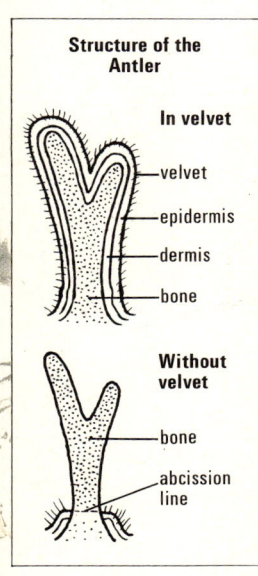

In velvet
- velvet
- epidermis
- dermis
- bone

Without velvet
- bone
- abcission line

Red Deer

female

97

FALLOW DEER
Cervus dama
Shoulder height about 100cm. Many colour varieties, but usually reddish-yellow above, with white spots; yellowish-white below. Greyer with less conspicuous spotting in winter. Rump with a bold black-and-white pattern and a black stripe down the long tail. Antlers are found in male only and have a characteristic *palmate* (flattened) blade. These are clear of velvet by end of August and shed in May.
Range and habitat: deciduous and mixed forests with scrub and clearings. A very common deer in parks, from which many have escaped, so original range is difficult to define.
Food and habits: mainly grasses or herbs and berries. In winter may eat bark of young trees. Forms mixed herds in winter; males leave herd in spring.
Breeding: ruts in autumn, and usually a single young born in spring.

SIKA DEER
Cervus nippon
Shoulder height 80–90cm. Slightly smaller than the Fallow deer. Colour buff-brown with faint spots in summer and a grey head; in winter a darker brown. Short white tail; rump patch white with a dark edge. Antlers have up to 4 *tines* (branches), and are clean of velvet in September and cast in April.
Range and habitat: originally broad-leaved woodland in eastern Asia; introduced to many parks in Europe, where escapes have given rise to many feral herds in several countries.
Food and habits: grazes and browses on young tree and shrub shoots. May strip bark, occasionally becoming a pest of forestry.
Breeding: ruts in September–October; single calf born May–June. May interbreed with Red deer.

SPOTTED DEER
Cervus axis
Shoulder height 90cm. White spots obvious throughout the year. Tail long and with dark central stripe. Rump patch does not have the black margin found in Fallow deer. Antlers are long and thin (not palmate) and have a maximum of 3 points.

Range and habitat: an Indian species introduced to a few localities in Europe and established in Istria (Yugoslavia). Lives in woodlands.

Food and habits: grazes and browses on young shoots. Habits similar to Fallow deer.

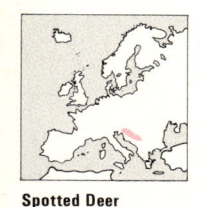

Spotted Deer

Spotted Deer

male, summer

male, summer

Sika Deer

Sika Deer

male, winter

99

REINDEER
Rangifer tarandus
Shoulder height 110–120cm. Coat thick, and variable in colour, but often a dark grey-brown in summer and paler in winter. Ears small, tail short. Hooves particularly broad and deeply cleft. Both sexes carry antlers with complex shape. Forward-pointing brow tine is branched. Antlers of female smaller than males, but carried through winter, whereas male sheds antlers in early winter.

Range and habitat: lives on the tundra and taiga of northern Europe. Domesticated in parts of northern Europe especially Lappland. Domestic animals small and very variable in colour.

Food and habits: grasses and sedges, and some small animals in summer; lichens and young shoots in winter. Wild herds are migratory, and strong swimmers.

Breeding: ruts in late autumn; a single calf or twins born in spring.

ELK
Alces alces
Shoulder height 1·8–2m. The largest deer. Colour grey-brown to black above, lighter on snout and legs. Legs very long. Broad muzzle with characteristic downward curve. Males carry large palmate antlers with numerous branches.

Range and habitat: once close to extinction, but herds are now expanding in many areas, especially southern

A reindeer calf can walk just two hours after birth and is weaned in two months. Nearly all European reindeer are domesticated and are cared for by herdsmen who follow the migrations north in the spring and south again in autumn. The domestic animals are small in size and variable in colour.

Elk

Sweden. Domesticated in parts of Russia. Lives in marshlands in summer, moving to drier land, often woodland in winter.
Food and habits: eats leaves and young shoots, particularly birch and willow, and aquatic plants in summer. In winter subsists on bark and shoots. Spends much time in or near water, and can swim well.
Breeding: ruts in September. Usually twins born in May or June.

WHITE-TAILED DEER
Odocoileus virginianus
Shoulder height 105cm. Coat reddish-brown in summer, greyish in winter, with no spots. Broad tail dark on upper side, almost hiding white rump patch except when raised in alarm. Fawns spotted.
Range and habitat: widespread in North America, and introduced to parts of Finland, where it has become established. A forest animal.
Food: mainly leaves, including the needles of conifers; also herbs and grasses.
Breeding: November rut; usually twin fawns born in summer.

ROE DEER
Capreolus capreolus
Shoulder height 75cm, female smaller. A small, slender deer. Summer coat smooth and red-brown in colour; winter coat long and grey. Underparts white. Tail almost invisible. Small, upright antlers with characteristic knobbed base (the *coronet*) and only 3 points. Antlers clear of velvet in May, shed in November. Fawns spotted.

male, winter

Roe Deer

Roe Deer

male, summer

fawn

Muntjac

Chinese Water Deer

Chinese Water Deer

Range and habitat: abundant in woodlands of northern Europe, less common in south. Prefers open woodland, but may be found in any region with enough cover.
Food: mainly a browser eating leaves of trees and shrubs, but will also take grasses, nuts, fungi and herbs. In winter eats shoots and may bark trees.
Breeding: mates in late summer; single young or twins born in early summer of following year.

MUNTJAC
Muntiacus reevesi
Shoulder height 40cm. A tiny deer with a dark red coat marked with white on chin, throat and rump. Coat lighter in female. Tail longish and bushy. Male carries simple antlers pointing backwards from prominent pedicels, which are long and hairy. Females carry tufts of hair in place of antlers. Fawns spotted.
Range and habitat: from southern China, and introduced to parks and estates in England and France, where escaped animals are now established. Lives in woodland with cover, and grazes in clearings.
Food and habits: both grazes and browses, eating shrubs, shoots, grass and fruit. May strip bark and cause crop damage.
Breeding: no clearly defined breeding season. A single fawn is born at any time of year.

CHINESE WATER DEER
Hydropotes inermis
Shoulder height 50cm. Very small. Pale brown coat, marked with black and white around the face. Large ears and small dark tail. No antlers in either sex. Male has long, curved canine teeth forming tusks (smaller in female). Fawns spotted.
Range and habitat: from eastern China, and now well established, through escapes from parks, in southern England. Lives in grassland, open woodland and marshes.
Food and habits: eats mainly grasses, but also some roots and vegetables. Solitary, nocturnal and not easy to find.
Breeding: ruts in December; young born May or June. Usually twins, but more than 2 fawns not uncommon.

WHALES, DOLPHINS AND PORPOISES
Order Cetacea

A distinctive order containing about 90 species of fully aquatic animals called cetaceans, which do not come ashore. The order is distributed worldwide, with about 25 species recorded in European waters, mostly in the sea, though a few are found in fresh water. They have forelimbs modified as flippers, no hind limbs, a streamlined shape, and a powerful tail with horizontal *flukes* to propel the animal through the water—all modifications for the aquatic way of life. Some species can remain under water for more than an hour. Nostrils open through the *blowhole* at the high point of the head.

Examples of Whale Baleen Plates

Blue whale

Black Right whale

Humpback whale

Krill

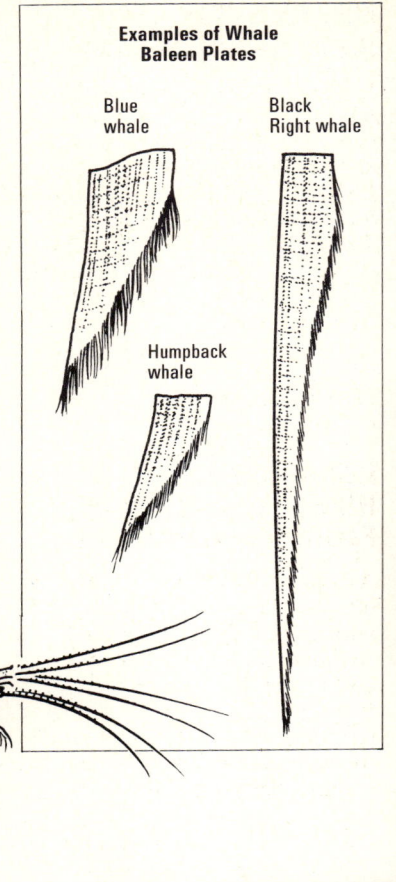

BALEEN WHALES
Sub-order Mystaceti

Large whales lacking teeth, which are replaced by large numbers of plates, known collectively as *baleen*. These are frayed along the inner edge and form a gigantic sieve to trap the food by filtering large amounts of water. The plates are made of a horny substance and vary in size and colour in the various species. The food of Baleen whales consists mainly of tiny prawn-like animals collectively called *krill*.

RORQUALS AND HUMPBACK WHALES
Family Balaenopteridae

A family of 6 species found in all oceans. They are fast swimmers, with a small dorsal fin placed well behind the middle of the body. The throat has many longitudinal grooves, and the flippers are long and pointed. All the whales in this family are commercially exploited.

BLUE WHALE
Balaenoptera musculus
Length up to 30m. The largest whale and the largest mammal ever to live. A large specimen can weigh over 100 tonnes. Dark blue-grey above and below, with pale flecks. Baleen black and up to 80cm in length. The blow is high and conspicuous.
Range and habitat: worldwide distribution, but rare due to exploitation and now protected. A few hundred survive in North Atlantic.
Food and habits: feeds on krill and small fish. Migratory, feeding in the cold northern waters in summer and moving south to warmer waters to breed in winter.
Breeding: a single young born every second year. Calves are about 2m long at birth.

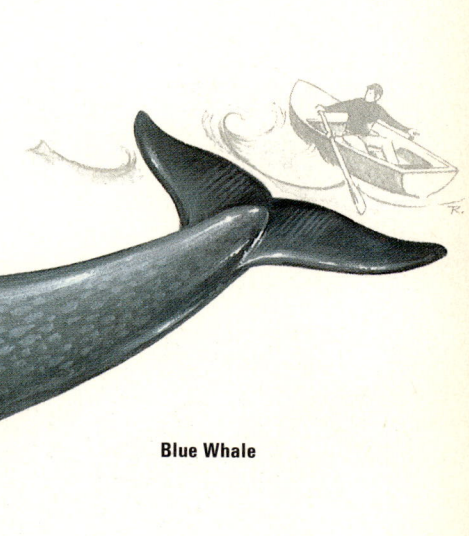

Blue Whale

FIN WHALE
Balaenoptera physalis
Length up to 20m. More slender than the Blue whale, and snout more pointed. Dark grey-black above, white below, but more white on right lower jaw than on left. The baleen is also coloured in this way, the front third of the plates on the right being white. The plates are up to 1m long. Blow is straight up, forming a conical shape.
Range and habitat: worldwide in temperate and polar waters.
Food and habits: feeds on krill and small fish. Gregarious, assembles in large herds and feeds near the surface.
Breeding: a single calf, occasionally twins, born every two years. Young born in warmer southern waters in winter.

SEI WHALE
Balaenoptera borealis
Length up to 18m. Smaller than the Fin whale, but more robust and with a larger dorsal fin with concave hind border. General colour blue-grey, lighter on the under-surface, often with a white throat patch. Baleen black with soft whitish fringes. Blow is low and not easily observed.
Range and habitat: worldwide in temperate and polar waters.
Food and habits: feeds on krill and even smaller crustaceans. Not normally gregarious, but seen alone or in pairs. Migrates towards Arctic waters in summer, and south to breed in winter.
Breeding: young born every second or third year.

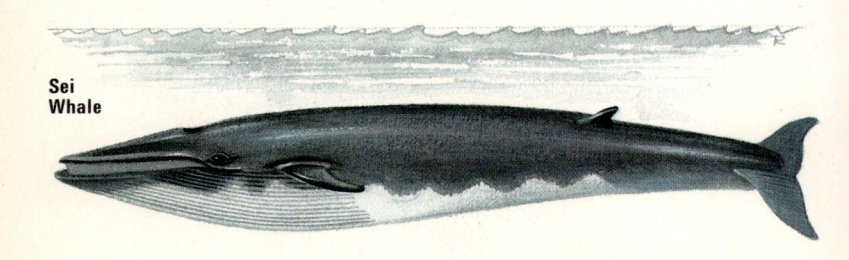

MINKE WHALE
Balaenoptera acutorostrata
Length up to 10m, the smallest Baleen whale in Atlantic waters. Robust. Blue-grey above, white below. White band or patch on large flippers is characteristic. Tail has a central notch. Baleen yellowish-white.
Range and habitat: worldwide in temperate and polar waters. Relatively common in European waters from Arctic to Mediterranean. Favours relatively shallow water near coasts.
Food and habits: feeds on small fish, squid and krill. Migrates north to feed in summer, south to breed in winter. Large collections of stones have been found in stomach.
Breeding: can produce a single calf each year.

HUMPBACK WHALE
Megaptera novaeangliae
Length up to 14m. A heavily built whale, the head being about a third of its total length. Black above, white below, including flippers and tail flukes, but colour pattern variable. Flukes irregularly indented behind, and with central notch. Dorsal fin small and set on a small hump. Baleen plates black, up to 1m long.
Range and habitat: worldwide, but in Atlantic has been hunted to near extinction.
Food and habits: feeds on plankton, krill and small shoaling fish. A slower, more stately swimmer than the other members of the family, and more common in shallow waters. Renowned for its haunting 'singing', which can be heard at great distances.
Breeding: 1 or, occasionally, 2 young born in summer.

RIGHT WHALES
Family Balaenidae

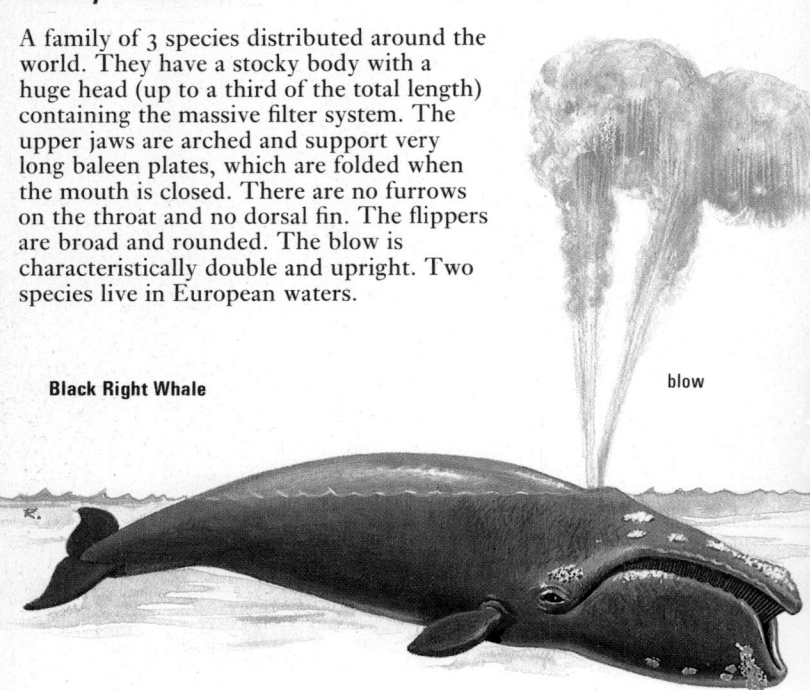

A family of 3 species distributed around the world. They have a stocky body with a huge head (up to a third of the total length) containing the massive filter system. The upper jaws are arched and support very long baleen plates, which are folded when the mouth is closed. There are no furrows on the throat and no dorsal fin. The flippers are broad and rounded. The blow is characteristically double and upright. Two species live in European waters.

Black Right Whale

blow

BLACK RIGHT WHALE
Balaena glacialis
Length up to 18m. Body black all over with a horny growth (the *bonnet*) on tip of upper jaw. Occasionally pale patches below. Often with a considerable growth of barnacles and other marine life. Head a quarter of total length. Baleen plates long (up to 3m), narrow and black.
Range and habitat: worldwide in temperate waters. Once common in North Atlantic, but was reduced to near extinction by whaling.
Food and habits: feeds on small crustaceans and krill. Gregarious. Migrates between Spanish coast and Arctic. A slow swimmer.
Breeding: a single young born in late summer.

BOWHEAD WHALE
Balaena mysticetus
Length up to 18m. Head even larger than that of Black Right whale, about a third of total length. Black with a cream-coloured chin region. Flippers short and broad. Baleen plates up to 3m long.
Range and habitat: Arctic regions of North America and Eurasia, but now relatively rare through over-exploitation by whalers.
Food and habits: similar to those of Black Right whale. Both species use lower jaw to scoop in water containing food, close mouth, raise massive tongue to expel water, and swallow food.
Breeding: usually a single young born in summer.

The characteristic 'bonnet' on the head of Right whales consists of horny material like hard skin and is often infested with worm and crustacean parasites.

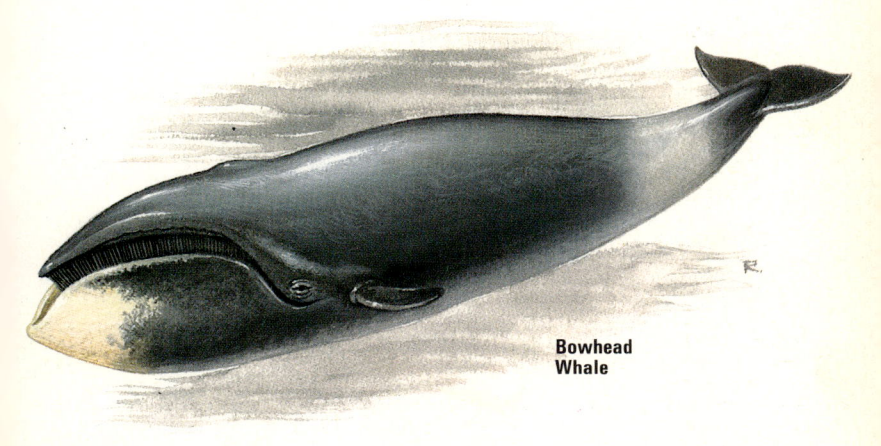

Bowhead Whale

TOOTHED WHALES
Sub-order Odontoceti

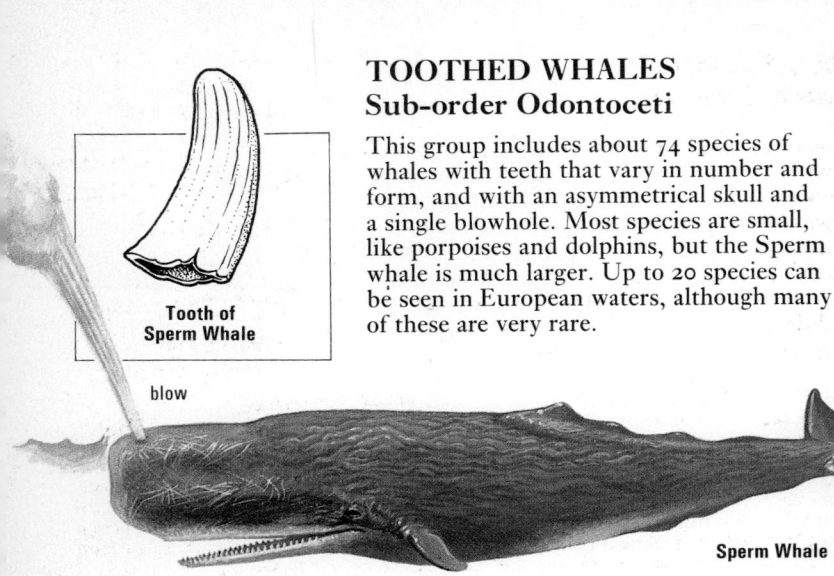

This group includes about 74 species of whales with teeth that vary in number and form, and with an asymmetrical skull and a single blowhole. Most species are small, like porpoises and dolphins, but the Sperm whale is much larger. Up to 20 species can be seen in European waters, although many of these are very rare.

Tooth of Sperm Whale

blow

Sperm Whale

SPERM WHALES
Family Physeteridae

The family includes 3 species, all worldwide in distribution. The head is large, with the teeth confined to the small lower jaw.

Pygmy Sperm Whale

SPERM WHALE
Physeter catodon
Largest of the Toothed whales. Length up to 18m. Huge rectangular head, about a third of total length. No dorsal fin, but a series of humps towards back of body. Lower jaw is short and narrow and carries many conical teeth which fit into sockets along the sides of the palate. Dark grey or black in colour, lighter on sides, grey or white below and some white on head. Flippers small. Head full of *spermaceti*, a waxy substance used in candles and ointments, and sperm oil, which is used as an industrial lubricant. Teeth used as ivory. *Ambergris*, from the intestines, is used to fix perfumes.
Range and habitat: worldwide, but mostly in warmer seas. Specimens in North Atlantic are mostly adult males.
Food and habits: feeds on fish, large squid and cuttlefish. Usually travels in family groups, but congregates into herds of hundreds. Males often solitary, and they alone venture into cold northern waters.
Breeding: single young born after a gestation period of about 16 months.

PYGMY SPERM WHALE
Kogia breviceps
Length up to 3·5m, females smaller. Small, with very short lower jaw which carries 12–16 conical teeth. Dorsal fin high and sickle-shaped.
Range and habitat: found in all oceans, but rare everywhere.
Food and habits: eats mainly squid and cuttlefish. Little known of habits.
Breeding: gestation period about 9 months.

BEAKED WHALES
Family Ziphiidae

About 14 species of this family are distributed around the world. Some are very rare. They are distinctly beaked, but smoother in profile than dolphins. They carry 1 or 2 pairs of enlarged teeth (*tusks*) in the lower jaws, though some females are toothless. They often carry scars caused by fighting with other members of the same species.

Tusk Positions in Beaked Whales

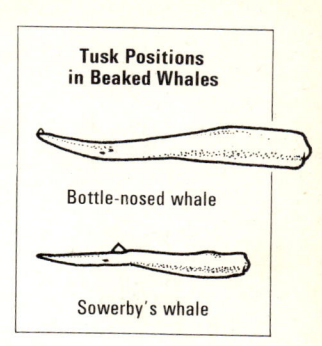

Bottle-nosed whale

Sowerby's whale

BOTTLE-NOSED WHALE
Hyperoodon ampullatus
Length up to 9m, females smaller. Snout of about 15cm, in front of bulbous forehead. Triangular fin. Dark grey to black in colour, lighter below and tending to become generally lighter in colour with age. Single pair of teeth at tip of lower jaw in males; also present, but do not develop, in female.
Range and habitat: North Atlantic, south to western African coast. Migrates to edge of ice in summer, south in winter.
Food and habits: eats squid, cuttlefish and herring. Habits little known, but sociable and found in schools of up to 50.
Breeding: gestation period about a year.

CUVIER'S WHALE
Ziphius cavirostris
Length up to 8·5m. Stoutly built. Forehead not bulbous. Colour variable, but usually with face and back light cream and the remainder black, or whole body may be greyish-buff. Males with single conical tooth at tip of each lower jaw; not visible in female.
Range: worldwide.
Food and habits: eats mainly squid and cuttlefish. Little known of habits, but seen in groups of 30–40.
Breeding: gestation period about a year.

Cuvier's Whale

Bottle-nosed Whale

Sowerby's Whale

True's Beaked Whale

SOWERBY'S WHALE
Mesoplodon bidens
Length up to 5m. Black above, with some white on belly, variable in extent. Single pair of teeth situated about halfway along lower jaw. In male, teeth are triangular and flattened; in female, they are smaller and may be concealed.
Range and habitat: restricted to North Atlantic; known from strandings on both sides of the ocean.
Food and habits: unknown.
Breeding: unknown.

TRUE'S BEAKED WHALE
Mesoplodon mirus
Length up to 5·2m. Dark grey above, grey on flanks, white below. Body slender and flattened from side to side. A single pair of teeth at extreme front of lower jaw, directed forwards and flattened in cross section. Teeth concealed in female.
Range and habitat: North Atlantic. A few specimens have stranded on European coasts and in northern America.
Food and habits: unknown.
Breeding: unknown.

GERVAIS' WHALE
Mesoplodon europaeus
Length up to 6m. Dark grey above, slightly lighter below. A single pair of teeth, flattened from side to side, are situated about one-sixth of length of jaw behind tip.
Range and habitat: North Atlantic, but only recorded from a single stranding.
Food and habits: unknown.
Breeding: unknown.

BLAINVILLE'S WHALE
Mesoplodon densirostris
Length up to 5·2m. Black above, grey below. Single pair of teeth in rear half of lower jaw. Jaw swollen to give a curving appearance to mouth.
Range and habitat: found in tropical and temperate seas, including southern North Atlantic.
Food and habits: unknown.
Breeding: unknown.

NOTE The *Mesoplodon* group of whales appear to spend most of their lives in deep water, seldom approaching coastal waters. They are very poorly known, and seldom seen.

Gervais' Whale

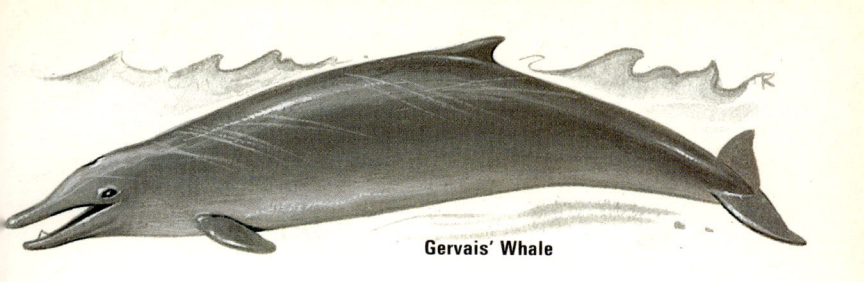

WHITE WHALES
Family Monodontidae

The family includes 2 species, found only in the North Atlantic. They lack a dorsal fin, but are otherwise similar to dolphins, with a blunt snout and no beak. The flippers are short.

WHITE WHALE
Delphinapterus leucas
Length up to 5m. Colour white or cream in adult. Has 8–10 pairs of simple conical teeth in both upper and lower jaws. Hunted for blubber and leather.
Range and habitat: an Arctic species occasionally found in temperate northern Atlantic.
Food and habits: eats cuttlefish, fish and crustaceans. Social in habits, but males form separate herds, except during breeding season.
Breeding: gestation period about 14 months, a single young being born in early summer every 2 or 3 years.

NARWHAL
Monodon monoceros
Length up to 5·5m. No dorsal fin. Pale in colour, but with dark flecks on back. Two teeth in upper jaw, one of which (generally the left, and usually in males only) develops into a long spiral tusk up to 3m long.
Range and habitat: Arctic seas, along coasts and occasionally up rivers.
Food and habits: eats cuttlefish, crustaceans and fish. Often seen in small groups, usually of same sex.
Breeding: little known.

White Whale

Narwhal

DOLPHINS
Family Delphinidae

This family includes about 55 species of small whales distributed around the world and with a few species living in fresh waters. Ten species can be found in European seas. They are fast swimmers, feeding mainly on fish. They usually have many small conical teeth. They often associate in groups, sometimes in schools of hundreds.

Long-finned Pilot Whale

Killer Whale

LONG-FINNED PILOT WHALE
Globicephala melaena
Length up to 8·5m. Bulbous forehead, similar to the Bottle-nosed whale, but with a very short beak. Black, except for white throat and chest. Dorsal fin large and recurved, about half-way down back. Long, slender, tapering flippers. Teeth 8–10 on each side of both jaws and arranged near front of jaw. Teeth are small, conical and about 10mm in diameter.

Range and habitat: the Atlantic and southern oceans. Frequently seen in coastal waters from Arctic to Mediterranean.
Food and habits: eats mainly cuttlefish and squid. Usually gathers in large schools of up to 100 animals. Migrates north in summer. Herded in Faroe Islands and driven ashore.
Breeding: long gestation period, so single young born every 2 or 3 years.

Bottle-nosed Dolphin

The Bottle-nosed dolphin is the most common dolphin in the North Atlantic and is often seen in shallow coastal waters. It is also the most common species trained for exhibition.

KILLER WHALE
Orcinus orca
Length up to 9·5m. Heavily built, with blunt snout and very high dorsal fin (up to 2m). Distinctive black-and-white colour pattern. Black above, mostly white below, with broad bands of white on sides and a patch behind eye. Teeth in upper and lower jaw number 10-13 on each side; teeth are oval in section.
Range: worldwide.
Food and habits: a fast, voracious carnivore, which eats other whales, seals, penguins and fish, especially salmon. May hunt in packs, especially to attack adult whales of large species. Can be tamed, and many are exhibited in dolphinaria, where they are docile and trainable.
Breeding: a single calf born in winter.

BOTTLE-NOSED DOLPHIN
Tursiops truncatus
Length up to 4m. A sleek, streamlined dolphin with a well-defined short beak. Black or grey-brown above, with a lighter to white belly. Dorsal fin is near middle of back and is concave behind. Teeth number 22-25 in each jaw and are conical and about 12mm in diameter.
Range and habitat: widely distributed, and is the commonest dolphin in European waters. Often seen in shallow water.
Food and habits: eats a wide variety of fish, depending on availability, and will also feed on shrimp and squid. Social in habits, and although often seen in a school of more than 100 animals, is usually found in smaller groups. This is the familiar dolphin seen in dolphinaria; it is tameable and can be taught to perform tricks. Also 'playful' in wild.
Breeding: gestation period about 1 year; a single young born during the spring.

RISSO'S DOLPHIN
Grampus griseus

Length up to 3·5m. Bulbous forehead, but no obvious beak. High, curved dorsal fin (up to 0·5m). Grey above, often marked with pale scars; lighter below, but amount of white very variable. Fins and tail black. Teeth number 3–7 on each side in lower jaw, but absent or occasionally only 1 or 2 in upper jaw.

Range: worldwide, but rare in far north of Atlantic.

Food and habits: feeds on squid and cuttlefish. Solitary, or seen in small groups. May follow ships and known to be 'playful', jumping well clear of the water. Scars on body are thought to be result of fighting with other members of the same species or of attacks by large squid.

Breeding: unknown.

ROUGH-TOOTHED DOLPHIN
Steno bredanensis

Length up to 2·5m. Upper parts blue-grey or purplish-black, with scattered markings of paler colour. Pinkish-white below, with darker streaks. Beak white and slender. Surfaces of teeth have vertical furrows and wrinkles. There are 20–27 teeth on each side of each jaw.

Range and habitat: found occasionally in warmer European waters, but normally tropical and not common north of the Mediterranean.

Food and habits: food unknown. Social in its habits and seen in schools of up to 50 animals. May follow ships.

Breeding: unknown.

Common Dolphin

COMMON DOLPHIN
Delphinus delphis
Length up to 2·4m. Beak long (about 10–12cm), slender and well defined by a distinct groove from forehead. Colour black above, white below, with a complex pattern of grey, yellow and white on sides. Two conspicuous waves of colour meet below dorsal fin. Dark circle around eye extending on to forehead. Teeth simple and about 3mm in diameter. There are 40–50 on each side of each jaw.
Range and habitat: worldwide in temperate and warm seas, but occasionally ventures into colder Arctic waters. Common in Black Sea, Mediterranean and northwards to southern Britain.
Food and habits: feeds mainly on fish, squid and cuttlefish. Gregarious in habits. A fast swimmer, reputed to reach 25 knots.
Breeding: gestation period about 10 months. A single young born in summer.

FALSE KILLER WHALE
Pseudorca crassidens
Length up to 5·5m, females smaller. Black all over and slender in shape, with a large dorsal fin but no beak. Teeth large in size, round in section, measuring about 25mm in diameter. There are usually 8–10 in each side of each jaw.
Range and habitat: worldwide, except in cold polar water. Usually a deep-sea dolphin.
Food and habits: eats fish and squid. Little known of habits, but travels in large groups. This species is known mostly through examining stranded animals.
Breeding: probably no definite season.

False Killer Whale

STRIPED DOLPHIN
Stenella coeruleoalba

Length up to 2·5m. Similar to Common dolphin in general proportions. Characteristic colour pattern consists of a number of narrow black lines extending back from eye region and branching. The general colour is black above, with a white belly and a complex pattern of greys on the sides but no yellow or brown pigment. Teeth about 3mm in diameter, and numbering 40–50 on each side of each jaw.

Range and habitat: worldwide in warm and temperate waters, the most northerly European record being southern Britain. Appears to prefer offshore deep water.

Food and habits: feeds mostly on fish and cuttlefish. Gregarious in habits, and often seen alongside ships.

Breeding: unknown.

Striped Dolphin

White-sided Dolphin

White-beaked Dolphin

Common Porpoise

WHITE-SIDED DOLPHIN
Lagenorhynchus acutus
Length up to 3m. A robust species with a distinct beak about 5cm long. Snout, head and back dark; below white. A narrow dark stripe runs from mouth to flipper, and there is a complex colour pattern on the sides. Hind part of body compressed from side to side, so that tail is deep and narrow just in front of the flukes. Teeth 30–40 on each side of each jaw.
Range: widely distributed in North Atlantic south to the North Sea.
Food and habits: eats fish, especially herring and cod. A gregarious animal found in large schools (up to 1,000 animals have been recorded in a school). Some southerly migration in winter. Can be trained in captivity.
Breeding: gestation period about 10 months; young born in spring.

WHITE-BEAKED DOLPHIN
Lagenorhynchus albirostris
Length up to 3m. The short white beak and the less-compressed body distinguish this species from the White-sided dolphin. Pale band on flank also extends further forwards, and the dorsal fin is larger. There are 22–25 teeth, about 6mm in diameter, on each side of each jaw.
Range: widely distributed in North Atlantic; very common in the North Sea, and can be found south to Portugal.
Food and habits: eats fish, including herring, cod and whiting. Habits similar to those of the White-sided dolphin.
Breeding: calves born in summer.

PORPOISES
Family Phocoenidae

The true porpoises are small animals, never measuring more than 2m in length, with flattened, spade-like teeth which may have lobed crowns. They have no beak, the snout being blunt. About 7 species are distributed around the world, of which only 1 occurs in European waters.

COMMON PORPOISE
Phocoena phocoena
Length up to 1·8m. The smallest cetacean. Stoutly built, without a beak and carrying a small, broad dorsal fin. Black above, white below. Flippers black. Colour can be variable, with some grey on the sides. Teeth ·22–27 on each side of each jaw, but flattened, not conical as in dolphins. In young animals the crowns may be lobed.
Range and habitat: widely distributed in North Atlantic and from White Sea to Mediterranean and Black Seas. Prefers coastal waters, and may enter estuaries and larger rivers.
Food and habits: eats fish, cuttlefish and crustaceans. Found in small groups. Relatively slow swimmers.
Breeding: gestation period about 1 year, with a single young born July–August.

GLOSSARY

baleen: plates of horny material in upper jaw of toothless whales, used to filter food from the water.
browser: an animal that feeds on tender young shoots and leaves of trees and shrubs.

calcar: a spur of cartilage or bone at the ankle of a bat which helps to support the tail membrane.
canine: a conical tooth situated between the incisors and the cheek-teeth, particularly prominent in carnivores.
carnassials: blade-like cheek-teeth of carnivores adapted for chopping and shearing flesh and bone.
cartilage: the flexible, gristly material which forms the skeleton in young animals and is later mostly transformed into bone.
cheek teeth: the premolars and molars, the teeth behind the canines (in rodents behind the diastema).
chromosome: material in the cell nucleus which carries the genes.
cusp: a projection on the surface of a tooth.

deciduous: describes a forest in which the trees shed their leaves seasonally.
dentine: the hard bony material which forms most of the substance of a tooth.
diastema: the space between teeth in a jaw.
digit: a finger or toe.
diurnal: describes an animal active mainly during the day.

enamel: the hard calcareous substance which covers the surface of a tooth.

feral: living in the wild; used to describe domesticated animals which are living independently in the wild.
fur: the soft, fine hairy part of the coat of a mammal.

gestation: the period in which the young develop in the uterus of the mother.
grazer: an animal which feeds on herbs and grasses.

guard hair: the long stiff hairs which project through the fur.

habitat: the environment in which an animal is usually found.
hibernation: a long, winter sleep in which the animal's temperature is lowered and its bodily functions are slowed.

incisors: front teeth, situated in front of the canines.

krill: shrimp-like crustaceans which live in large, dense shoals and are used as food by baleen whales.

leveret: a young hare in the first year of life.
litter: the offspring of a single birth.

molars: the posterior cheek teeth, not preceded by 'milk teeth'.

nocturnal: describes an animal active mainly at night.

premolars: the cheek teeth between the canines and molars which are preceded by 'milk teeth'.
prey: an animal taken by a predator as food.

sella: central part of the nose leaf in horse-shoe bats, below the lancet.
species: a group of similar animals which can freely interbreed in the wild.
steppe: dry, flat, usually treeless grassland.

tragus: a fleshy lobe growing up from the lower part of the ear in bats.
tusk: an enlarged tooth, usually a canine (in narwhal an incisor).

velvet: soft hairy material which covers and nourishes growing antlers and is rubbed off when antlers are fully formed.

INDEX

**Mammals
(English Names)**

Agouti 34
Antelope 89, 90
Ape 15; Barbary 15
Ass 88
Auroch 91

Baboon 15
Badger 70, 77
Barbastelle 30, 31
Bat 10, 11, 22–31;
  Bechstein's 27; Blasius's
  Horse-shoe 24, 25;
  Brandt's 26, 27; Common
  Long-eared 30; Common
  Pipistrelle 22;
  Daubenton's 26; Free-
  tailed 22, 31; Fruit 22;
  Geoffroy's 27; Greater
  Horse-shoe 24, 25;
  Greater Mouse-eared 28;
  Greater Noctule 28; Grey
  Long-eared 30; Horse-
  shoe 24–5; Kuhl's
  Pipistrelle 30; Leisler's 28,
  29; Lesser Horse-shoe 24;
  Lesser Mouse-eared 28;
  Lesser Noctule 28; Long-
  fingered 26;
  Mediterranean Horse-
  shoe 24, 25; Mehely's
  Horse-shoe 24, 25;
  Myotis 26–8; Nathaline
  26; Nathusius's Pipistrelle
  30; Natterer's 26, 27;
  Noctule 26, 28, 29;
  Northern 29; Parti-
  coloured 29; Pipistrelle
  22, 26, 30; Pond 26;
  Savi's Pipistrelle 30;
  Schreiber's 30; Serotine
  26, 28, 29; Vespertilionid
  22, 26–30; Whiskered 27.
  See also Barbastelle
Bear 64, 65; Brown 65;
  Polar 64, 65
Beaver 34, 40; Canadian 40;
  European 40
Birch Mouse 63; Northern
  63; Southern 63
Bison 10, 90, 91
Boar, Wild 89
Buffalo 90

Capybara 34
Carnivore 11, 13, 64–87
Cat 32, 79, 80–1; Domestic
  80, 81; Wild 9, 33, 80, 81
Cattle, Domestic 89, 90, 91
Cetacean 104, 119
Chamois 94, 95

Chipmunk, Siberian 39
Civet 78
Coypu 42

Deer 12, 67, 75, 80, 89,
  96–103; Chinese Water
  103; Fallow 98, 99; Red
  10, 96, 97; Roe 10,
  102–3; Sika 98, 99;
  Spotted 99; White-tailed
  102
Desman 16; Pyrenean 17
Dingo 66
Dog 12, 32, 64, 66–9
Dolphin 12, 104, 110, 113,
  114–19; Bottle-nosed
  115; Common 117, 118;
  Risso's 116; Rough-
  toothed 116; Striped 118;
  White-beaked 118, 119
  White-sided 118, 119
Dormouse 43–5; Common
  44, 45; Fat 44; Forest 44;
  Garden 43; Mouse-tailed
  45

Elk 67, 100–1

Ferret, Domestic 73
Fox 10, 32, 33, 38, 64;
  Arctic 10, 68, 69; Blue 68;
  Red 68

Genet 64, 78, 79
Giraffe 89
Glutton 75
Goat 67, 90, 92; Domestic
  94; Wild 94

Hamster 46–7, 72; Common
  46; Golden 47; Grey 47;
  Rumanian 47
Hare 10, 14, 32, 80; Brown
  33; Mountain 33
Hedgehog 10, 16; Algerian
  16; Eastern 16; Western
  16
Herbivore 89
Hooved Mammals 11; Even-
  toed 89–103; Odd-toed
  88
Horse 12, 88; Camargue
  88; Przewalski 88
Horse-shoe Bat 24–5;
  Blasius's 24, 25; Greater
  24, 25; Lesser 24;
  Mediterranean 24, 25;
  Mehely's 24, 25

Ibex 94; Alpine 92, 93;
  Spanish 93
Insectivore 11, 17–21

Jackal 10, 66, 67

Kangaroo 14
Kinkajou 79

Lemming 10, 48, 68;
  Norway 48; Wood 48
Lemur 12
Lynx 80; Spanish 80

Marmot, Alpine 38, 39
Marsupial 14
Marten 36; Beech 35, 74,
  75; Pine 10, 35, 74, 75
Mink 70; American 72;
  European 72; Ranch 72
Mole 12, 16; Blind 17;
  Common 10; Northern 17;
  Roman 17
Mole-rat 55; Greater 55;
  Lesser 55
Mongoose 64, 78–9;
  Egyptian 78, 79; Indian
  Grey 79
Monkey 15
Mouflon 92
Mouse 10, 46, 56, 58–63,
  71, 75; Algerian 62; Birch
  63; Cretan Spiny 62;
  Harvest 61; House 62;
  Northern Birch 63; Pygmy
  Field 58; Rock 60;
  Southern Birch 63;
  Steppe 62; Striped Field
  60; Wood 58, 60; Yellow-
  necked Field 58
Muntjac 103
Muskrat 54, 55
Mustelid 64

Narwhal 113

Otter 70, 72, 76
Ox, Musk 9

Panda, Greater 79;
  Lesser 79
Pig 11, 89
Pika 32
Pine Vole 52–3; Alpine 52;
  Bavarian 53; Common 52;
  Liechtenstein's 53;
  Lusitanian 53;
  Mediterranean 53; Savi's
  53; Tatra 53; Thomas's 53
Pipistrelle 22, 26, 30;
  Common 22; Kuhl's 30;
  Nathusius's 30; Savi's 30
Polecat 10, 70; Marbled 73;
  Steppe 72, 73; Western
  72, 73
Pony, Exmoor 88; New
  Forest 88
Porcupine 41; North African
  Crested 41
Porpoise 104, 110, 119;
  Common 119
Primate 15
Procyonid 64

121

Rabbit 10, 32, 68, 71, 72, 73, 74, 76, 80
Raccoon 64, 79
Raccoon-dog 69
Rat 12, 46, 56–7, 60; Black 57, Common 56, 57; Norway 56; Ship 57
Reindeer 9, 11, 67, 100, 101
Rhinoceros 88
Rodent 10, 11, 34–63, 69, 71, 72, 74, 75, 78, 79, 80
Rorqual 105

Sable 10
Seal 11, 82–5, 115; Bearded 85; Common 82; Grey 83; Harp 84; Hooded 85; Monk 84; Ringed 82, 83
Sea Lion 82
Sheep, Domestic 67, 89, 90, 92; Soay 92
Shrew 10, 11, 12, 16, 18, 75; Alpine 19; Apennine 19; Bicoloured White-toothed 21; Common 18, 19, 21; Dusky 19; Etruscan 21; Greater White-toothed 21; Laxmann's 19; Least 19; Lesser White-toothed 21; Miller's Water 20; Millet's 18; Pygmy 18, 19; Pygmy White-toothed 21; Red-toothed 18; Spanish 19; Water 20; White-toothed 18
Souslik 72; European 38; Spotted 38
Squirrel 34–9, 43, 68, 74; Flying 34, 36; Gliding 34; Grey 36; Ground 34, 38; Red 10, 34, 35
Stoat 13, 64, 71, 72

Tapir 88
Tiger 64

Viverrid 64
Vole 10, 46, 47, 48–54, 68, 71, 76; Alpine Pine 52; Balkan Snow 49; Bank 48, 49; Bavarian Pine 53; Cabrera's 51; Common 50, 51; Common Pine 52; Field 50, 51, 52, 54; Grey-sided 49; Günther's 50, 51; Leichtenstein's Pine 53; Lusitanian Pine 53; Mediterranean Pine 53; Northern Red-backed 49; Northern Water 54; Pine 52–3; Savi's Pine 53; Sibling 50, 51; Snow 50, 51; Southern Water 54; Tatra Pine 53; Thomas's Pine 53

Wallaby 12, 14; Red-necked 14
Walrus 82, 86
Water Buffalo, Domestic 91
Water Vole 72; Northern 54; Southern 54
Weasel 13, 32, 64, 70, 71, 75
Whale 9, 12, 104–19; Baleen 105–9; Beaked 111–12; Black Right 108; Blainville's 112; Blue 105, 106; Bottle-nosed 111, 114; Bowhead 108, 109; Cuvier's 111; False Killer 117; Fin 106; Gervais' 112, 113; Humpback 105–7; Killer 114, 115; Long-finned Pilot 114; Minke 107; Pygmy Sperm 110; Right 108–9; Sei 106; Sowerby's 112; Sperm 110; Toothed 110–19; True's Beaked 112; White 113
Wolf 10, 64, 66, 67

Mammals (Scientific Names)

Acomys minous 62
Alces alces 100–1
Alopex lagopus 68
Apodemus agrarius 60; *flavicollis* 58; *microps* 58; *mystacinus* 60; *sylvaticus* 58
Artiodactyla 89
Arvicola sapidus 54; *terrestris* 54

Balaena glacialis 108; *mysticetus* 108
Balaenidae 108
Balaenoptera acutorostrata 107; *borealis* 106; *musculus* 105; *physalis* 106
Balaenopteridae 105
Barbastella barbastellus 30
Bison bonasus 91
Bos taurus 91
Bovidae 90
Bubalus bubalus 91

Canidae 64, 66
Canis aureus 67; *lupus* 67
Capra aegagrus 94; *hircus* 94; *ibex* 92; *pyrenaica* 93
Capreolus capreolus 102
Capromyidae 42
Carnivora 64
Castor canadensis 40; *fiber* 40
Castoridae 40
Cercopithecidae 15

Cervidae 96
Cervus axis 99; *dama* 98; *elaphus* 96; *nippon* 98
Cetacea 104
Chiroptera 22
Citellus citellus 38; *suslicus* 38
Clethrionomys glareolus 49; *rofocanus* 49; *rutilus* 49
Cricetinae 46
Cricetulus migratorius 47
Cricetus cricetus 46
Crocidura leucodon 21; *russula* 21; *suaveolens* 21
Crystophora cristata 85

Delphinapterus leucas 113
Delphinidae 114
Delphinus delphis 117
Dinaromys bogdanovi 49
Dryomys nitedula 44

Eliomys quercinus 43
Eptesicus nilssoni 29; *serotinus* 29
Equidae 88
Equus asinus 88; *caballus* 88
Erignathus barbatus 85
Erinaceidae 16
Erinaceus algirus 16; *concolor* 16; *europaeus* 16

Felidae 80
Felis catus 81; *lynx* 80; *sylvestris* 80

Galemys pyrenaicus 17
Genetta genetta 79
Glis glis 44
Globicephala melaena 114
Grampus griseus 116
Gulo gulo 75

Halichoerus grypus 83
Herpestes edwardsi 79; *ichneumon* 78
Hydropotes inermis 103
Hyperoodon ampullatus 111
Hystricidae 41
Hystrix cristata 41

Insectivora 13, 17–21

Kogia breviceps 110

Lagenorhynchus acutus 119; *albirostris* 119
Lagomorpha 32
Lemmus lemmus 48
Leporidae 32
Lepus capensis 33; *timidus* 33
Lutra lutra 76

Macaca sylvanus 15
Macropodidae 14
Macropus rufogriseus 14
Marmota marmota 38
Marsupialia 14
Martes foina 75; *martes* 74
Megachiroptera 22
Megaptera novaeangliae 107
Meles meles 77
Mesocricetus auratus 47; *newtoni* 47
Mesoplodon bidens 112; *densirostris* 112; *europaeus* 112; *mirus* 112
Microchiroptera 22
Micromys minutus 61
Microtus agrestis 50; *arvalis* 51; *cabrerae* 51; *epiroticus* 51; *guentheri* 51; *nivalis* 51; *oeconomus* 51
Miniopterus schreibersi 30
Molossidae 22, 31
Monachus monachus 84
Monodon monoceros 113
Monodontidae 113
Muntiacus reevesi 103
Muridae 46
Murinae 46, 56
Mus hortulanus 62; *musculus* 62; *spretus* 62
Muscardinus avellanarius 44
Mustela eversmanni 72; *erminea* 71; *furo* 73; *lutreola* 72; *nivalis* 71; *putorius* 72; *vison* 72
Mustelidae 64, 70
Myocastor coypus 42
Myomimus roachi 45
Myopus schisticolor 48
Myotis bechsteini 27; *blythi* 28; *brandti* 26; *capaccinii* 26; *dasycneme* 26; *daubentoni* 26; *emarginatus* 27; *myotis* 28; *mystacinus* 27; *nathalinae* 26; *nattereri* 27
Mystaceti 105

Neomys anomalus 20; *fodiens* 20
Nyctalus lasiopterus 28; *leisleri* 28; *noctula* 28
Nyctereutes procyonoides 69

Odobenidae 86
Odocoileus virginianus 102
Odontoceti 110
Ondatra zibethicus 54
Orcinus orca 115
Oryctolagus cuniculus 32
Ovibos moschatus 91
Ovis aries 92; *musimon* 92

Pagophilus groenlandicus 84
Perissodactyla 88
Phoca hispida 82; *vitulina* 82
Phocidae 82
Phocoena phocoena 119
Physeter catodon 110
Pinnipedia 82
Pipistrellus kuhli 30; *nathusii* 30; *pipistrellus* 30; *savii* 30
Pitymys bavaricus 53; *duodecimostatus* 53; *leichtensteini* 53; *lusitanicus* 53; *multiplex* 52; *savii* 53; *subterraneus* 52; *tatricus* 53; *thomasi* 53
Plecotus auritus 30; *austriacus* 30
Primates 114
Procyonidae 64
Pseudorca crassidens 117
Pteromys volans 36

Rangifer tarandus 100
Rattus norvegicus 56; *rattus* 57
Rhinolophidae 22, 24
Rhinolophus blasii 24; *euryale* 24; *ferrumequinum* 24; *hipposideros* 24; *mehelyi* 24
Rodentia 34
Rupicapra rupicapra 94

Sciuridae 34
Sciurus carolinensis 36; *vulgaris* 35
Sicista betulina 63; *subtilis* 63
Sorex alpinus 19; *araneus* 18; *caecutiens* 19; *coronatus* 18; *granarius* 19; *minutissimus* 19; *minutus* 19; *samniticus* 19; *sinalis* 19
Soricidae 16, 18
Spalacinae 55
Spalax leucodon 55; *microphthalmus* 55
Stenella coeruleoalba 118
Steno bredanensis 116
Suidae 89
Suncus etruscus 21
Sus scrofa 89

Tadarida teniotis 31
Talpa caeca 13, 17; *europaea* 13. 17; *romana* 13, 17
Talpidae 13, 16, 17
Tamias sibiricus 39
Thalarctos maritimus 65
Tursiops truncatus 115

Ursidae 64, 65
Ursus arctos 65

Vespertilio murinus 29
Vespertilionidae 22, 26
Viverridae 64, 78
Vormela peregusna 73
Vulpes vulpes 68

Zapodidae 63
Ziphiidae 111
Ziphius cavirostris 111

ACKNOWLEDGEMENTS

The author and publishers wish to thank the following for their help in supplying photographs for this book on the pages indicated:

Bruce Coleman 109 (Jen & Des Bartlett); Coypu Research Laboratory, Ministry of Agriculture, Fisheries and Food 42; Marineland of Florida 115; P. Morris 53; Natural Science Photos 101 (Geoffrey Kinns); Nature Photographers Ltd 23 (S. C. Bisserot), 20 (Kevin Carlson), cover, 37, 45 (Owen Newman); NHPA 81 (R. Balharry), 59 (Stephen Dalton), 61 (Roger Hosking), 87 (Stephen J. Krasemann); ZEFA 8, 47, 68, 76 (H. Reinhard).

Picture research by Penny Warn.
Maps by Sally Brooks